By the end of August 1944 the German army in the West had been dealt a devastating blow. Apart from the tanks of 9.Panzer-Division, which had joined the battle after the withdrawal from Normandy had begun, and 11.Panzer-Division, which managed to stay one step ahead of the enemy, Oberbefehlshaber West (OB West), the overall commander of German forces on the Western Front, could call on little more than a handful of armoured vehicles.

During the first weeks of September the Allies were pushing towards the Rhine and many observers were optimistic that the war could be brought to a successful conclusion by Christmas. But within weeks the front had been stabilised, largely due to the efforts of local commanders, and during October and November the armoured units which had been almost completely destroyed in the fighting in France were rebuilt.

As early as mid-December 1944 the Germans were capable of mounting a major offensive operation in the Ardennes region and, in January 1945, a large-scale attack to recapture Strasbourg. While these actions were ultimately costly failures they did signal that the conflict was far from over.

In this book, the third in the TankCraft series to deal with the Pzkpfw IV, we look at the last production models of this famous armoured vehicle and the units which operated it in the final months of the war. Readers will notice that most of the photographs in this book depict vehicles which are either disabled or abandoned. This can be attributed in some measure to Hitler's promise, following the defeat in Normandy, that the bulk of Panther and Tiger tank production would be diverted to the Western Front. At the same time he insisted that the Pzkpfw IV chassis be utilised to carry larger and more powerful guns and at one point halted production of the Pzkpfw IV ausf J at the Nibelungenwerk in favour of the new Jagdpanzer designs (1). Krupp-Grusenwerk estimated that the introduction of new tanks and the almost constant amendments to production targets instigated between April and December 1943 resulted in the loss of over 300,000 man hours. In addition, the disastrous losses suffered in the summer 1944 battles in Byelorussia, Ukraine and eastern Poland had to be made good and many of the Pzkpfw IV tanks assembled in the later months of the war were dispatched to units fighting in the East.

Importantly, this was a time when Germans needed to be reassured that the war, if not winnable, could be brought to an acceptable conclusion and one way of doing this was to highlight the Wehrmacht's technical superiority. Consequently, the larger, more powerful tanks feature heavily in photographs and newsreel footage. The enduring image of the Ardennes Offensive is that of the Tiger II tanks of Kampfgruppe Peiper rolling unchallenged through Belgian towns.

But the Pzkpfw IV remained an important weapon in the arsenal of the Panzertruppen and would take its place in what would prove to be the war's deadliest phase.

Notes

1. It was planned that by November 1944 the Pzkpfw IV was to be replaced by Panzer IV/70(A) and Panzerjäger III/IV tank destroyers. The Panzer III/IV project, which included the latter, and the Panther II, which was to eventually replace all these types, constituted an attempt to introduce more effective weapons while easing manufacture by rationalising many of the component parts. But these fantasies resulted in significant holdups at all the factories and only the Nibelungenwerk was involved in Pzkpfw IV production from June 1944.

A Pzkpfw IV of II.Abteilung, Panzer-Regiment 16 photographed during the Ardennes Offensive. The regiment's tanks were concentrated in a Kampfgruppe led by Oberst Johannes Bayer which also included parts of Panzergrenadier-Regiment 60, Panzerartillerie-Regiment 146 and a company from Panzer-Pionier-Bataillon 675. The regiment was authorised two full companies of Pzkpfw IV tanks but only sixteen were available on 16 December 1944 when the assault began and fourteen of these had been lost by Christmas Eve.

Our map shows the situation in the West from the beginning of 1945 to May of the same year, when the last German units surrendered. Note that the national boundaries are not those of the 1940s, where Austria had been incorporated into the Reich and western Czechoslovakia was a German protectorate known as Bohemia and Moravia, but rather those with which present-day readers will be more familiar. Changes to the front line and lost German territory are shown as follows: A. The front as of Friday, 5 January 1945 when it had been stabilised somewhat following the Ardennes Offensive, which principally affected the area between Monschau and Trier, and Operation Nordwind, where most of the fighting took place betweem Sarreguemines and Strasbourg. B. The Allied advance up to 24 March 1945, following the Rhine crossings. C. The approximate boundaries of the Ruhr Pocket on 5 April 1945, the day after US 1st and 9th Armies completed the encirclement of a large part of Heeresgruppe B. D. The front line on 18 April 1945. E. The furthest extent of the Allied advance as of 8 May 1945 when hostilities in the West officially ended. Note that, in places, this exceeded the boundaries of the demarcation line agreed to with the Soviets, particulary in southern Germany and Czechoslovakia. Other locations mentioned in the text but not listed by name above include: 1. Panzertruppen-Schule Bergen. 2. Schiess-Schule der Panzertruppen Putlos. 3. Lehrtruppe der Panzertruppen-Schule Fallingbostel. 4. Truppenübungsplatz Sennelager. 5. Truppenübungsplatz Ordruf. 6. Truppenübungsplatz Alten-Grabow. 7. Truppenübungsplatz Grafenwohr. 8. Sankt Valentin, where the Nibelungenwerk was located. 9. Plauen, the site of the VOMAG-Maschinenfabrik AG plant. 10. Panzer-Stützpunkt Nord. 11. Panzer-Stützpunkt Mitte. 12. Panzer-Stützpunkt Süd 1944. 13. Panzer-Stützpunkt Mayen. 14. The small village of Bad Kötzting where the last tanks of Panzer-Regiment 15 surrendered on 4 May 1945.

Listed here are the units which took part in the final battles on the Western Front, principally the defence of German towns and cities, and were equipped with the Pzkpfw IV tank. Much of the information regarding allocations and unit strengths was gleaned from the official Führungsliste, or allocation and shipping lists, and this is explained further in the charts on pages 14 and 15.

Throughout this section, as a matter of convenience, I have used the abbreviations PzBefWg for Panzerbefehlswagen, the purpose-built command tank, and of course Pzkpfw for Panzerkampfwagen. A small number of Bergepanzer IV, recovery tanks built on the Pzkpfw IV chassis, were produced from October 1944 and I have also included these where accurate information is available. Although allocations of the Panzerbeobachtungswagen IV, the specialist artillery observation tank, are shown in the charts mentioned above I have not included the artillery regiments which operated them in the following list as they were not under the control of the Generalinspekteur der Panzertruppen. Those units will be covered in a future title. The Heereszeugämter (HZA) mentioned here were the various ordnance depots from which vehicles were shipped to operational units, training establishments or arsenals.

Panzer-Regiment 3. Formed in October 1935 from Reiter-Regiment 12, the regiment was attached to 2.Panzer-Division and remained with that formation until the end of the war. The division fought in the Normandy battles and the subsequent retreat through France where it was largely destroyed.

By November the Panzer regiment had been rebuilt and reported on 14 December 1944, just prior to the commencement of the Ardennes Offensive, that II.Abteilung was equipped with twenty-two Sturmgeschütz III assault guns, with 5.Kompanie and 6.Kompanie, and twenty-four Pzkpfw IV tanks assigned to 7.Kompanie and 8.Kompanie. The regiment's I.Abteilung had sixty-four Panthers.

For the main attack the division was divided into four Kampfgruppen and the tanks of II.Abteilung, with most of the regimental headquarters, supported Panzerjäger-Abteilung 38 under the command of Hauptmann Friedrich Holtmeyer.

In January 1945, with the conclusion of the Ardennes Offensive, the division retreated towards Clervaux in northern Luxembourg and in the following weeks fought in the defensive battles in the Eifel around Prüm and Bitburg and along the Moselle River. On 15 March the commander of II.Abteilung reported that although sixteen Pzkpfw IV tanks and twenty-two assault guns were on hand, just a handful of these vehicles were considered operational (1).

In early April 1945, the division was operating in the area around Schweinfurt and Bamberg and, badly depleted, was promised the second battalion of Panzer-Regiment 22 as reinforcement but these tanks never reached their destination and instead supported Kampfgruppe von Hobe fighting further to the south near Munich (2). In these last battles the commander of the regiment's headquarters company, Oberleutnant Peter Prien was nominated for the award of the Ritterkreuz on two separate occasions but the war ended before he could receive the medal.

Notes

1. The regiment also had thirty-five Panthers, although just four of these were considered to be combat-ready. While the remnants of Panzerausbildungs-Verband Thüringen, an ad-hoc unit formed from elements of Panzertruppen-Schule Bergen, were attached to the division in late March 1945, they probably brought little in the way of armoured assets with them.
2. The equipment, organisation and history of Kampfgruppe von Hobe is beyond the remit of this book but is is examined in *TankCraft 18: Panther Tanks, German Army and Waffen-SS, Defence of the West, 1945.*

The brief caption of this US Signal Corps photograph gives the date as 24 February 1945 and the location as Belgium. The US official history states that these vehicles were abandoned after confronting units of the US 3rd Army and both the OKW and US Army situation maps show that 2.Panzer-Division was the only armoured unit on 3rd Army's front at this time. This Pzkpfw IV ausf J is fitted with three return rollers, a modification that was introduced in October 1944, and parts of the supporting rail for the Drathgeflechtschürzen mesh screens and Flammentöter exhaust mufflers can still be seen. Note that the very neatly applied dots of the camouflage scheme do not extend to the turret Schurzen.

Panzer-Regiment 15. Formed in October 1937 from elements of Panzer-Regiment 2 and Panzer-Regiment 5, this unit took part in the fighting in Poland and the advance into Belgium and Flanders in 1940. Following the French campaign the regiment was attached to the newly formed 11.Panzer-Division.

When the Normandy invasion took place in June 1944 the division was stationed in southern France and when the Allied armies landed on the beaches of the Côte d'Azur on 15 August 1944, conducted a fighting withdrawal along the Rhône valley towards Besançon. At that time Panzer-Regiment 15, commanded by Oberst Meinrad von Lauchert, was at about half strength with twenty-six Pzkpfw IV tanks and forty-nine Panthers. The division also had a handful of self-propelled guns with Panzerjäger-Abteilung 51.

Although the fighting here was just as fierce as the battles taking place further to the north, the Germans had continually managed to escape encirclement, unlike the units in Normandy. Consequently, by September 1944, 11.Panzer-Division was the only armoured formation in the West which could field a full Panzer regiment made up of two battalions although the headquarters and all the companies were understrength (1).

The division fought in the battles in Alsace and took part in the Ardennes Offensive with II.Abteilung reporting that thirty-one Pzkpfw IV tanks were on hand with a further thirty in transit, although eleven of the latter did not arrive until 26 and 27 December by which time the offensive was almost over.

During the first months of the new year the division took part in the battles in the Saarland and along the Moselle River and later in the defence of Cologne and the attacks on the Remagen bridgehead near Bad Honnef in support of 9.Panzer-Division. But by this time most German formations were divisions in name only and Panzer-Regiment 15 reported on 15 March 1945 that just seventeen Pzkpfw IV tanks were on hand, of which four were fully serviceable.

In addition to being understrength, the tanks of Panzer-Regiment 15 had been forced to disengage from their positions around Dusseldorf and travel to Remagen on their own tracks, a distance of over 100 kilometres, at a time when spare parts were in short supply. In April the division was transferred to 7.Armee based in central Germany and withdrew through Hesse and Thuringia towards the Sudetenland. By the end of the month the division was strung out between Domazlice and Volary in the present-day Czech Republic and had been split into a Kampfgruppe, made up of a company of tanks with the infantrymen of Panzergrenadier-Regiment 111, and the main body of the division and both were effectively immobilised by a complete lack of fuel.

On 2 May 1945, the division received orders to move east to face the advancing Red Army but realising that his men could not possibly halt the Soviet advance their commander, Generalleutnant von Wietersheim (2), approached the US Army units to the west and offered his division's surrender which took effect two days later. At this time the Panzer regiment could field a single Panther and six Pzkpfw IV tanks and some of these needed to be towed to the American lines.

At this point some mention should be made of the the unusual numbering system adopted by II.Abteilung in the last weeks of the war.

During the battles of 1944 and early 1945 the Pzkpfw IV tanks of Panzer-Regiment 15 were identified by a three-digit company number which was carried on the sides and, usually, the rear of the turret Schürzen. The style of these numbers varied considerably, even within the same company, but they followed the official system which most readers will be familiar with that denoted the company, the platoon and the vehicle's place within the platoon (3). For example, the number 631 would identify the first tank of the sixth company's third platoon.

Photographs of the battalion's Pzkpfw IV tanks disabled during the retreat through France show that the three-number system was still in use then. The regiment operated near Trier and then Remagen until 23 March but was by that time reduced to a handful of vehicles. An allocation of twelve new Pzkpfw IV tanks, dispatched directly from Nachschub West, was made on 17 March 1945 but unfortunately no delivery date was recorded. At some time, possibly in early April, a two-digit identification system was adopted and examples are shown in the photographs on the facing page.

I can offer no explanation as to the exact significance of the numbers nor give a more precise date for their introduction. The losses incurred in March could certainly have necessitated some reorganisation. It may be that the first digit indicated the platoon, numbered consecutively throughout the battalion, while the second identified the individual vehicle but this is purely conjecture on my part. Interestingly, a similar system was employed in the early stages of the fighting on the Eastern Front. Some numbers were rendered neatly, to a standard size, but others were quite rough and their application obviously hurried as can be seen from our photographs.

........................*text continued on page 6*

Notes

1. Hermann Balck, who commanded Heeresgruppe G from September to December 1944, rated Panzer-Regiment 15 as one of the best units of the German Army and attributed the regiment's ability to keep its tanks going to the efforts of Oberfeldwebel Kurt Reuschel of the maintenance section. Wietersheim, the division commander, credited Major Karl Heinz Voigtmann, his logistics officer, for keeping all the supply elements just far enough to the rear to be withdrawn in time.
2. Wietersheim, who had led the division through the battles in France and Germany, had been promoted to a command in Berlin and replaced by Generalmajor Freiherr Treusch von Buttlar-Brandenfels. Feigning illness he remained with the division and negotiated its surrender.
3. A number of the regiment's Pzkpfw IV tanks from this period are shown in *TankCraft 35: Panzer IV Medium Tank, German Army and Waffen-SS, Normandy Campaign, Summer 1944.*

Examples of the two-digit numbering systen used by Panzer-Regiment 15 in the last weeks of the war. 1. Pzkpfw IV ausf J photographed in May 1945, possibly near Bad Kötzting, east of Regensburg near the present-day Czech border, where the last of the regiment's vehicles were handed over. The number 11, partially visible here, is confirmed by other images from the series. 2. Pzkpfw IV ausf J photographed at a vehicle dump, possibly Pilsen, after May 1945. The number 13, with the Balkenkreuz just visible, can be seen on the turret Schürzen side. 3. Pzkpfw IV ausf J number 21 photographed on 4 May 1945 when the bulk of 11.Panzer-Division surrendered to US Army units at Bad Kötzting. Most accounts agree that just six Pzkpfw IV tanks and a single Panther were present on the day. A film made of the event shows number 21, number 25 and at least two Pzkpfw IV tanks without numbers. All were painted in the same factory-applied camouflage scheme. 4. Pzkpfw IV ausf J number 22. The background of this image would seem to suggest that it was photographed at the same location as number 13. 5. Pzkpfw IV ausf J number 23. Although I have been unable to positively identify this vehicle it would seem to have been photographed at the same location as number 22. 6. Pzkpfw IV ausf J number 25. This vehicle was photographed with Pzkpfw IV number 21, shown above, at Bad Kötzting on 4 May 1945. 7. Pzkpfw IV ausf J. The number 72 would seem to support the contention that the system identified the platoon and individual tank. I should mention that the date given for this photograph is the subject of some debate. Not shown on this page, but featured in the Camouflage & Markings section, is Pzkpfw IV number 33. The placement of the Balkenkreuz on that tank and the style and colour of the number is identical to number 72.

.........................text continued from page 4

Panzer-Regiment 16. Originally raised as a Panzer battalion for 16.Infanterie-Division (mot.) in mid-1942, by May 1944 this unit had been reorganised as a full regiment with two battalions (1). Attached to 116.Panzer-Division the regiment took part in the battles in Normandy, including Operation Lüttich, and was almost completely destroyed in the Falaise Pocket.

The regiment was rebuilt in November 1944 with I.Abteilung equipped with Panthers and II.Abteilung organised as a mixed battalion with two companies of Pzkpfw IV tanks and one company of Sturmgeschütz III assault guns. However, at the start of the Ardennes Offensive, less than one month later, the regiment reported that just forty-one Panthers and sixteen Pzkpfw IV tanks were available for immediate deployment (2) while the assault guns were still in transit.

During February 1945 the division supported the paratroopers of 1.Fallschirmarmee, fighting along the Niederrhein, and in March opposed the Allied crossings of the Rhine, managing to take part in the defence against both the land and airborne phases of Operation Varsity.

In the report of 15 March 1945, II.Abteilung reported that six Pzkpfw IV tanks and eleven assault guns were on hand but only two of each type was serviceable. But within days twenty Pzkpfw IV tanks were shipped from the HZA to reinforce the regiment and these arrived on 29 March.

For much of April 1945 the mass of the division was subordinated to Armee-Abteilung-Lüttwitz and was largely destroyed in the battles for the Ruhr Pocket, surrendering to units of the US Army on Wednesday, 18 April 1945. A small Kampfgruppe, under the direct command of OB West, fought on the Elbe and in the Harz Mountains until the end of the month when their fuel and ammunition were exhausted.

Panzer-Regiment 22. Raised in May 1943, the regiment was attached to the reformed 21.Panzer-Division and fought in the Normandy battles where it lost most of its equipment. At the end of July 1944, II.Abteilung was withdrawn from the front to begin training on the Panther (3).

In late September 1944 Panzer-Brigade 112 was disbanded and handed over its remaining vehicles, including twenty-five Pzkpfw IV tanks and ten Panthers, to the regiment's I.Abteilung which was with 21.Panzer-Division near Èpinal as part of 5.Panzerarmee.

In October the battalion was organised as a mixed Panzer-Abteilung with 1.Kompanie and 2.Kompanie equipped with Panthers while 3.Kompanie and 4.Kompanie were outfitted with Pzkpfw IV tanks. On 14 November 1944 ten Pzkpfw IV tanks were shipped to the battalion and the Führungsliste notes that this brought the total dispatched to twenty-nine. In the following month another two were sent and on 29 December the battalion reported that eighteen Pzkpfw IV tanks were available for combat while another fifteen tanks and a PzBefWg IV were in repair.

Notes

1. Somewhat confusingly, the unit's original title, Panzer-Regiment 116, was changed to Panzer-Regiment 16 on 23 May 1944 after it joined 116.Panzer-Division.
2. This number is taken from the division's Kriegstagbuch and the figure of nineteen given in a number of accounts probably includes three Flakpanzer IV. During the first day of the operation the regiment lost six Pzkpfw IV tanks.
3. The battalion never returned to its parent formation and its history is related in *TankCraft 18: Panther Tanks, German Army and Waffen-SS, Defence of the West, 1945.*

Photographed in the town of Benniwihr, just north of Colmar in eastern France, this Pzkpfw IV ausf J was probably assembled in January 1945 as evidenced by the Lost Tafel poison gas detector panel on the gun mantlet. It may lack the air intakes on the hull brake access hatches, indicating a February-production model, but these are obscured by the spare track links. The OKW situation maps place Panzer-Regiment 22 and SS-Panzer-Regiment 10 in this area and other tanks from the latter were similarly devoid of any markings including company numbers.

Photographed in March 1945 this Pzkpfw IV ausf J of 3.Kompanie, Panzer-Regiment 33 may be one of the tanks lost in the fighting near Langlire, north-east of Houffalize in early January. Note the supporting rail for the Drathgeflechtschurzen mesh screens, one of which is still in place. Note also the swiveling hatch of the commander's cupola. The Prinz Eugen unit insignia, also shown inset at left, was carried by most of the regiment's Pzkpfw IV tanks in the position shown here.

The division, with I.Abteilung, remained in the Saarpfalz region, around Saarlautern, taking part in the attack towards Strasbourg and the fighting in northern Alsace, until 1 February 1945 when it was transferred to the East. As part of 9.Armee the division was caught in the Halbe Pocket, south of Berlin, in May 1945. Although all the heavy equipment was lost many of the men were able to make their way to the west where they surrendered to the Americans.

Panzer-Regiment 33. Originally formed by renaming the Austrian Panzerwagen-Bataillon when the Österreichischen Bundesheer was incorporated into the Wehrmacht, from 1940 this formation was attached to 9.Panzer-Division. In March 1943 it was officially redesignated Panzer-Regiment Prinz Eugen (1).

The division did not take part in the Normandy battles until 10 August 1944 but was almost immediately caught in the Allied advance in the fighting around Alençon and in little more than a week had been reduced to the strength of a reinforced brigade.

By September 1944 the division, with a single tank company, was deployed south-west of Aachen and although greatly understrength, was pushed into the counterattack at Elst, Near Arnhem, and also the fighting around Venlo on the Eindhoven to Duisburg highway.

In November 1944 the regiment was rebuilt with I.Abteilung made up of 1.Kompanie and 2.Kompanie equipped with Pzkpfw IV tanks (2) and 3.Kompanie with Sturmgeschütz III assault guns, while II.Abteilung contained four companies of Panthers.

The division took part in the Ardennes Offensive, where it suffered very heavy casualties, and in the subsequent fighting in the Eifel mountain range. In late February 1945, 9.Panzer-Division was involved in the attacks against the Remagen bridgehead but failed to reduce the American perimeter, ending the battle with just fifteen serviceable tanks. On 15 March 1945 the battalion reported that five Pzkpfw IV tanks were on hand but just one of these was considered combat-ready.

In the same month the division was part of the German force defending Cologne but was pushed out of the city, also losing its commanding officer Generalmajor Harald Freiherr von Elverfeldt, and eventually forced into the Ruhr Pocket.

On 7 April 1945 an order was issued directing that the remnants of 9.Panzer-Division were to be formed into a Kampfgruppe consisting of one Panzer battalion made up of one company of Pzkpfw IV tanks and two companies of Panthers. The formation process was to be completed within four days and each company was to contain ten tanks. In addition, II.Abteilung of Panzer-Regiment 33 was to be rebuilt from various units including a Panzer company of Kampfgruppe 116.Panzer-Division (3).

On 9 April 1945 a total of six Pzkpfw IV tanks that had been returned to the HZA from repair were dispatched to the division, one of the last shipments of the war. But how much of the planned reorganisation was completed is not known as OB West ordered on 26 April that 9.Panzer-Division was to be formally disbanded and its personnel dispersed to other units.

Notes

1. Not to be confused with the Waffen-SS division of the same name.
2. Twenty-eight operational vehicles in total.
3. The new battalion was to contain a company of Jagdpanzer 38 tank destroyers, one company of Sturmgeschütz III assault guns and one company equipped with the Panzer IV/70/(V).

Panzer-Regiment 55. Raised from Reserve-Panzer-Regiment 5 in late February 1945, this unit was attached to 233.Reserve-Panzer-Division (1) a formation that was intend to be part of the Atlantik Wall defences along the west coast of Denmark.

In April 1945 many of the regiment's tanks were returned to Germany be used to form Panzer-Division Holstein although a number of Pzkpfw III and Pzkpfw IV tanks were retained, three of the latter armed with the long-barrelled 7.5cm L/43 gun. The regiment remained in Denmark until the end of the war, never seeing combat, and was used to train tank crews.

Panzer-Lehr-Regiment 130. Formed in January 1943, the regiment was attached to Panzer-Lehr-Division. The regiment's II.Abteilung contained four companies all equipped with Pzkpfw IV tanks. When the Normandy landings began I.Abteilung was in the process of converting to the Panther and its place in the division was filled by I.Abteilung, Panzer-Regiment 6 (2).

On 20 August 1944, after the fighting in Normandy and the subsequent retreat, the division was withdrawn from the front to be rebuilt and it was anticipated that the Panther battalion would rejoin its parent unit. But I.Abteilung was assigned to Panzer-Brigade 113 and the regiment's II.Abteilung was rebuilt as a mixed unit with 5.Kompanie and 7.Kompanie each allocated fifteen Panther tanks while 6.Kompanie and 8.Kompanie were equipped with a total of thirty-four Pzkpfw IV vehicles which were delivered between 27 October and 8 November 1944.

The Panzer regiment would retain this organisation throughout the fighting in Lorraine and later in the Ardennes and the battles along the lower Rhine as I.Abteilung did not return until 18 February 1945 and even then its remaining tanks supported 116.Panzer-Division in the defensive actions between Kalkar and Goch, on the Dutch border south of Cleve.

By 15 March 1945, the division's effective strength had been reduced to twenty-nine Panthers and just six Pzkpfw IV tanks and later that month, the surviving vehicles were concentrated in I.Abteilung while the Panzer crews without vehicles and the men of the maintenance company were formed into a new II.Abteilung and sent to the Panzertruppen-Schule at Bergen.

In the final weeks of the war the Panzer-Lehr-Division took part in the battles for the Remagen bridgehead and the fighting in the Ruhr Pocket where the last survivors surrendered to units of the US Army on 15 April 1945.

Panzer-Abteilung 103. This unit was attached to 3.Panzergrenadier-Division and served in Italy until mid-1944 when the division was transferred to western France.

The battalion had been converted to a Panzer-Sturmgeshütz-Abteilung in July 1943 made up of three companies equipped with Sturmgeshütz III assault guns. Most sources state that the battalion headquarters operated two Panzerbefehlswagen IV command tanks but the Führungsliste gives only one which was delivered on 25 August 1944.

Notes

1. The reserve designation was dropped from the division's title at the same time.
2. The rather complicated history of the regiment's Panther Abteilung is detailed in *TankCraft 3: Panther Tanks, German Army and Waffen-SS, Normandy Campaign 1944.*

Photographed in early 1945, this early-production Pzkpfw IV ausf J, identified by the spare track link bracket on the hull side, is probably one of very few vehicles finished with a coating of Zimmerit, discontinued in September 1944, and painted in one of the factory-applied camouflage schemes, introduced just weeks previously. Note the hooks attached to the trackguard to support the hull Schürzen which would have been the so-called Series I version (see page 56). These very early features suggest that this tank of 8.Kompanie, Panzer-Lehr-Regiment 130 may be one of the vehicles allocated to the regiment's II.Abteilung in July 1944.

Panzer-Abteilung 115. Formed in late 1943, the battalion was attached to 15.Panzergrenadier-Division. In July 1944 the division, which had been fighting on the Italian Front, was transferred to France where it was later engaged in the battles in Lorraine and the fighting for the West Wall along the Saar valley. A report of 1 October 1944 states that the battalion had twenty-one Panzer IV tanks, presumably brought from Italy, and twenty-seven Jagdpanzer IV L/48 tank destroyers on hand. The same report mentions two Panzerbefehlswagen III command tanks but the relevant Führungsliste suggests that these vehicles may have in fact been Panzerbefehlswagen IV. In the same month the surviving assets of Panzer-Brigade 113 were absorbed by the division. Although the battalion was allocated thirty Sturmgeshütz III assault guns in the first week of December 1944 no further tank deliveries were made before the end of the war.

The division took part in the Ardennes Offensive, with nine serviceable Pzkpfw IV tanks, and in early 1945 was involved in the defensive battles between Roermond and Kleve including the counterattack on the Wesel bridgehead north of Duisburg. The last elements of 15.Panzergrenadier-Division surrendered to British army units near Lamstedt on 5 May 1945.

Panzer-Brigade 106. The brigade's Panzer-Abteilung 2016 was made up of three companies of Panther tanks with one company equipped with Panzer IV L/70(V) tank destroyers and in addition two Pzkpfw IV tanks were reported as being on hand on 8 December 1944. The latter were still with the battalion on 15 January 1945 but they are not mentioned when the remnants of Panzer-Abteilung 2016 joined Panzer-Regiment Clausewitz in early April 1945 (1).

Führer-Begleit-Brigade. The origins of this unit are somewhat complicated and many accounts of its later history rely on the report compiled in May 1947, from memory, by the brigade's commander, Ernst Otto Remer, while a prisoner of the Americans (2).

It is sufficient to record here that in April 1943 two combat units were formed from Hitler's security detail and named Führer-Begleit-Bataillon (FBBtl) and Führer-Grenadier-Bataillon (FGBtl). In January 1944, the FBBtl was sent to the Eastern Front where it served in the Baltic states and East Prussia and was eventually withdrawn in August.

In the first weeks of September, with Oberst Remer now in command, the battalion was expanded to regimental-size but retained just a single company of tanks referred to as 5.(Pz.) Kompanie, FBBtl. Remer's account also mentions a Sturmgeschütz company but I have been unable to find any allocation of assault guns to the brigade before February 1945. On 27 September 1944 a total of seventeen Pzkpfw IV tanks were shipped from the HZA and these arrived two day later. On 10 October five Panzer IV/70(A) tank destroyers were sent to the battalion and these may constitute the assault gun company that Remer recalled.

At the end of November the regiment was officially renamed Führer-Begleit-Brigade (FBB) and at the about the same time was reinforced with Sturmgeschütz-Brigade 200 and II.Abteilung, Panzer-Regiment Grossdeutschland (GD). The latter had been allocated seven Pzkpfw IV tanks, which were dispatched on 29 November 1944 and arrived on 2 December, and thirty-eight Panzer IV/70(A) tank destroyers which were with the battalion two days later.

The assault gun battalion was assigned twenty-two Sturmgeschütz III vehicles and nine Sturmhaubitze self-propelled howitzers. There is some suggestion that the attachment of these two units was seen as a temporary measure at the time. In addition, two Grenadier battalions were transferred in from Führer-Grenadier-Brigade (FGB) and a further three Pzkpfw IV tanks arrived on 27 December while the brigade was sheltering in the Bois de Herbaimont, north-west of Bastogne.

The brigade's organisation in the days leading up to the commencement of the Ardennes Offensive is unclear but Remer's account states that the tanks and assault guns were formed into an ad-hoc regiment of two battalions with I.Abteilung made up of four companies all equipped with Pzkpfw IV tanks. This would require sixty-two vehicles in total and this could have been achieved with the Pzkpfw IV and Panzer IV/70(A) tank destroyers received up to that time. It should be remembered that the official November 1944 Panzer-Regiment organisation, shown in detail on Page 61, allowed for mixed units of these vehicles. He also states that II.Abteilung contained four batteries of assault guns.

A detailed table of organisation published by an authoritative source states that on 16 December 1944, the day the Ardennes Offensive began, the first, second and third companies of the brigade's I.Abteilung were equipped with a total of thirteen, twelve and ten Panzer IV/70(A) tank destroyers respectively with the remaining three tank destroyers assigned to the battalion headquarters.

The battalion's 4.Kompanie contained seventeen Pzkpfw IV tank, five Sturmgeschütz III assault guns and five Sturmhaubitze self-propelled howitzers. The late Thomas Jentz in *Panzertruppen* volume 2 records that one company of the brigade had seventeen Pzkpfw IV tanks and five Panzer IV/70(A) tank destroyers and this arrangement seems more plausible. Neither, however, takes into

Notes

1. There is no record of any Pzkpfw IV allocation in the official Führungsliste and these two tanks may have been taken over from an unknown unit. They are mentioned as being attached to a Kampfgruppe led by Hauptmann Paul te Heesen, the commander of Panzer-Abteilung 2016.

2. Oberst, later Generalmajor, Remer was instrumental in thwarting the 20 July plot to kill Hitler as many readers will probably know. He was given command of the brigade as a reward for his loyalty.

This early-production Pzkpfw IV ausf J of Panzer-Abteilung 115 was abandoned in late December 1944 in the town of Sâvi, today an outer suburb of Bastogne. In the original print traces of Zimmerit are clearly visible on the hull side and this, with the early hull towing brackets, would suggest that this tank was assembled in Septmber 1944. This vehicle is also shown and discussed in the Camouflage & Markings section.

account the seven Pzkpfw IV tanks shipped to Panzer-Regiment GD in November. Most sources agree that the three batteries of Sturmgeschütz-Brigade 200 operated as the battalion's II.Abteilung. Although Remer mentions that the assault gun battalion was made up of four batteries, he is almost certainly incorrect in this.

The brigade supported 18.Volksgrenadier-Division in the attacks around Bastogne until the end of December 1944 by which time 4.Kompanie had lost five Pzkpfw IV tanks as total write-offs.

When the brigade was expanded to a division in late January 1945 the tanks and tank destroyers were used to form II.Abteiling of the newly-created Panzer-Regiment 102 and a fresh I.Abteilung was formed with Pzkpfw V Panther tanks. Sturmgeschütz-Brigade 200 was renamed Panzerjäger-Abteilung 673 and according to the division's last surviving Kriegsgliederung, dated 1 March 1945, was subordinated to Panzer-Regiment 102.

In February 1945 the division was transferred to the Eastern Front where it was largely destroyed in the fighting around Spremberg.

Führer-Grenadier-Brigade. In early November 1944 the brigade's armour component, referred to as III.Abteilung, was made up of three companies of Pzkpfw V Panther tanks and a company of Panzer IV/70(V) tank destroyers. The brigade had been allocated seven Pzkpfw IV tanks which were shipped from the HZA on 5 July 1944 and one authoritative source states that they were still on hand in the following December but I have been unable to determine their fate. Several accounts mention that Heeres-Sturmartillerie-Brigade 911, attached to the FGB from 16 December 1944, had seven or eight Pzkpfw IV tanks although I can find no mention of these in any official allocation list.

Schwere Panzer-Abteilung 424. Badly depleted by the fighting on the Eastern Front, this battalion was disbanded in February 1945 with the remnants of the first and second companies forming schwere Panzerjäger-Abteilung 512, one of only two Jagdtiger battalions (1).

In March and April the crews of 3.Kompanie, which had been training at Paderborn under the command of Oberleutnant Busch, were incorporated into a Kampfgruppe made up from the tanks of Panzer-Ersatz-und Ausbildungs-Abteilung 500 and allocated two Tiger I tanks, a single Panther and one Pzkpfw IV tank. One of the Tigers was destroyed in the fighting at Paderborn railway station but much of the Kampfgruppe was able to reach Höxter, between Paderborn and Northeim, where they eventually surrendered to the Allies.

Schwere Panzerjäger-Abteilung 559. This battalion was equipped with Jagdpanther tank destroyers, Sturmgeschutz III assault guns and, from December 1944, Panzer IV/70(V) tank destroyers. One source mentions three PzBefWg IV on hand in December but I have been unable to find any reference in either the Pzkpfw IV or the Befehls und Beobachtungswagen lists.

Notes

1. These units are examined in *TankCraft 42: Jagdtiger Heavy Tank Destroyer, German Army Western Front, 1945.*

Panzerjäger-Abteilung 1212. Attached to 212.Volksgrenadier-Division, the battalion is mentioned in at least one reliable source as having a Bergepanzer IV on hand in January 1945 although there is no record of any allocation in the Führungsliste. The division was built around 212.Infanterie-Division and that formation had been assigned a single Bergepanzer III in July 1944 but it is unlikely that any of the heavy equipment survived the summer battles in the East.

Panzerjäger-Abteilung 1276. Attached to 276.Volksgrenadier-Division, the battalion is reported to had operated a single Bergepanzer IV in January 1945 but there is no record of any allocation in the Führungsliste. The battalion's 3.Kompanie was equipped with Sturmgeschütz IV assault guns and it is possible that a recovery vehicle was handed over from one of the training establishments.

Panzerjäger-Abteilung 1559. Attached to 559.Volksgrenadier-Division, the battalion was allocated a single Bergepanzer IV which was shipped on 4 January 1945. It was presumably assigned to the battalion's 3.Kompanie which was referred to as Sturmgeschütz-Abteilung 1559 (1).

Panzer-Abteilung Putlos. This unit was created on 12 April 1945, from elements of Schiess-Schule Putlos, the German army's tank gunnery school, and attached to Panzer-Division Clausewitz. The battalion contained two mixed Panzer companies and one Panzergrenadier company. The seven Pzkpfw IV tanks on hand were all concentrated in 2.Kompanie. On the afternoon of April 17, after being reinforced with two Panther tanks and a number of armoured halftracks, the battalion was reorganised and renamed Kampfgruppe von Bennigsen.

Panzer-Kompanie Kummersdorf. This battle group was formed on 31 March 1945 and contained four Panthers, one Pzkpfw III and two Pzkpfw IV tanks with one Hummel and one Nashorn self-propelled gun. In addition, the company had a Tiger I and one Tiger II and a single Jagdtiger. On 19 April 1945, Panzer-Kompanie Kummersdorf was at Luchau near Hamburg and may have seen some action here against the advancing British. On 21 April the company was ordered to join Kampfgruppe Möws south of Berlin. This unit should not be confused with Panzer-Abteilung Kummersdorf which was absorbed by Panzer-Division Müncheberg and fought in Berlin.

Panzer-Stützpunkt Mayen. Forward support bases, or Stützpunkte, were normally set up at major rail or road junctions and their function was to administer the supply of vehicles, spare parts and other necessary materials. In early 1945 the support bases in the West included Panzer-Stützpunkt Nord based at Bergisch Gladbach, outside Cologne, Panzer-Stützpunkt Mitte, centred on Kaiserslautern and Panzer-Stützpunkt Süd, which had been relocated to Mühlacker near Pforzheim after the base at Hagenau had been abandoned.

During January 1945 the establishment located at Mayen, approximately 20 kilometres west of Koblenz, and Panzer-Stützpunkt Mitte were both allocated seventeen Pzkpfw IV tanks.

The ultimate fate of these vehicles is unknown although the Mayen vehicles certainly did not take part in the defence of the town. The history of the US 11th Armored Division, which captured Mayen on 8 March 1945, states clearly that the German defences were overcome by dismounted infantry who faced only small arms fire and the occasional Panzerfaust.

Panzer-Stützpunkt Mitte. See Panzer-Stützpunkt Mayen.

Many formations of the Waffen-SS, particularly those listed here, underwent a number of reorganisations and name changes at various times throughout the war. As these names will be familiar to most readers, and can be somewhat cumbersome, I have referred to the divisions and regiments by their numbers only.

SS-Panzer-Regiment 1. Formed in October 1942, the regiment was attached to 1.SS-Panzer-Division, the premier formation of the Waffen-SS.

The regiment served in Russia and Normandy where it was largely destroyed and by 3 September 1944 was withdrawn from the front to be rebuilt. Due to a shortage of armoured vehicles SS-Panzer-Regiment 1 was reorganised as a Gemischte, or mixed, Panzer-Abteilung, a single battalion made up of 1.Kompanie and 2.Kompanie with Panthers and 6.Kompanie and 7.Kompanie, each equipped with seventeen Pzkpfw IV tanks.

The battalion famously took part in the Ardennes Offensive as Kampfgruppe Peiper which was tasked with seizing the Meuse bridges at Huy and Amay, south-west of Liège. Peiper's unit managed to reach Stavelot and Trois-Ponts, almost halfway to their objectives, but were held up by stiff resistance and were eventually surrounded in Stoumont, where the tanks ran out of fuel.

Although most of the men were able to escape, all the vehicles were abandoned. On 10 January 1945, the division was withdrawn to the area east of St Vith and then to Cologne where the Panzer battalion was rebuilt. In February 1945 the division was transferred to the East.

Notes

1. The somewhat confusing nomenclature of these units is explained fulling in *TankCraft 19: STUG III & IV, German Army, Waffen-SS and Luftwaffe Units, Western Front 1944-1945.*

Photographed after the fighting in the Ardennes, this Pzkpfw IV ausf J of 7.Kompanie, SS-Panzer-Regiment 1 is fitted with the Flammentöter exhaust mufflers and the support rails and brackets for the Drathgeflechtschürzen, both introduced into production in September. Note that the mesh screen is still in place on the right side. The placement of the company number is typical and it was also displayed on the forward section of the turret Schürzen.

SS-Panzer-Regiment 2. Raised in October 1942 and attached to 2.SS-Panzer-Division, the regiment served in Normandy and took part in the fighting around St Lô, at Mortain and in the attempted relief of the Falaise Pocket. On 28 August the division was able to report that six serviceable tanks of all types were all that remained.

The regiment was rebuilt in November 1944 with I.Abteilung made up of four companies equipped with the Panther while II.Abteilung was allocated two companies of Pzkpfw IV tanks and two companies of Sturmgeschütz III assault guns. The division took part in the Ardennes Offensive, initially as part of the 5.Panzerarmee reserve with 9.SS-Panzer-Division, and was later engaged in the heavy fighting around Baraque de Fraiture, north of Houffalize, at Grandmeril and near St. Vith. Just prior to the beginning of the offensive II.Abteilung reported that twenty-eight Pzkpfw IV tanks and twenty-eight assault guns were on hand.

On 17 January 1945, with the offensive completely stalled, the division was pulled back to the town of Prüm, behind the Westwall defences, and in February, transferred to the East as part of II.SS-Panzerkorps. At about this time the division reported that a single Bergepanzer IV was on hand but there is no record of any allocation in the relevant Führungliste.

SS-Panzer-Regiment 9. Formed in February 1943, the regiment was attached to 9.SS-Panzer-Division and took part in the fighting in Normandy, including the costly battle for Mont Ormel. Any surviving tanks were handed over to 11.Panzer-Division in September (1).

The Panzer regiment was rebuilt in November 1944 with I.Abteilung consisting of four companies equipped with Panthers while II.Abteilung was made up of two companies of Pzkpfw IV tanks and two companies of Sturmgeschütz III assault guns.

The division took part in the Ardennes Offensive as part of 6.Panzerarmee and fought in the battles to take Losheim and St Vith and later supported elements of Kampfgruppe Peiper in the attacks on La Gleize.

This action was the high-water mark of the assault and by 6 January 1945 the division was being pushed back from the area around Bastogne and by the end of the month had been withdrawn from the front line.

On 9 February 1945, the division began moving to the East and ended the war fighting in Hungary as part of 2.Panzerarmee.

SS-Panzer-Regiment 10. Originally formed in February 1943, the regiment was attached to 10.SS-Panzer-Division. During the fighting in Normandy and the subsequent retreat, the regiment lost all its tanks and was eventually rebuilt as a Gemischte Panzer-Abteilung with two companies of Pzkpfw IV and two companies of Panther tanks, reporting on 10 December 1944 that thirty-six of the former, one a Panzerbefehlswagen command tank, and ten of the latter were on hand.

Notes

1. Elements of the division famously took part in the defence of Arnhem later the same month.

Under the command of Sturmbannführer Ernst-Johann Tetsch, the battalion took part in Operation Nordwind, the last major offensive in the West, initially seeing action in the Gambsheim bridgehead and near Brumath, just north of Strasbourg. In the battles around Herrlisheim, which began on 16 January 1945, the battalion succeeded in destroying more than twenty-eight US Army Sherman tanks and capturing a further ten (1).

A supplementary attack was mounted from Schirrheim towards Hagenau on 24 January but was this eventually beaten back by intense artillery fire. The division was withdrawn from the front on 3 February 1945, reporting that eighteen combat-ready Pzkpfw IV tanks were on hand, and shortly after this began moving to the East to join Heeresgruppe Weichsel in Pomerania.

SS-Panzer-Regiment 12. Formed in November 1943, the regiment was attached to 12.SS-Panzer-Division and fought in Normandy as part of I.SS-Panzerkorps, losing all its heavy equipment in the battles around Falaise and the retreat to the Seine.

On 8 September 1944 all surviving elements of the division, with the exception of a small Kampfgruppe formed around SS-Panzergrenadier-Regiment 26, were ordered back to Germany.

In November the regiment's first battalion was rebuilt as a Gemischte, or mixed, Panzer-Abteilung with two companies of Pzkpfw IV and two companies of Panther tanks and this single-battalion organisation would remain in place until the end of the war. The Panther companies were numbered as 1 and 3 while the Pzkpfw IV companies were 5 and 6. On 8 December the battalion reported that a total of thirty-seven Pzkpfw IV tanks were on hand.

The division formed part of 6.Panzerarmee during the Ardennes Offensive, reporting that forty-one Panthers and thirty-seven Pzkpfw IV tanks were on hand just before the assault commenced. The Panzer regiment was badly mauled in the attack on Krinkelt and in the fighting further to the west around Büllingen and Bütgenbach.

In late December the division was subordinated to II.SS-Panzerkorps and was engaged in the battle for the Ourthe River bridgehead and in the fighting around Bastogne. On 10 January 1945 12.SS.Panzer-Division was withdrawn from the front and sent to Cologne.

On 2 February the division was transferred to the East and fought in Hungary as part of 8.Armee and 6.Panzerarmee, gradually withdrawing into Austria.

Notes

1. Sources differ on the total number captured but a report of 25 March 1945 gives the figure shown here. Interestingly, they remained on the division's inventory and were later used on the Eastern Front with the last surviving vehicle being destroyed by its own crew on the night of 7 May 1945.

Although few production features are visible here I have included this image as it shows a rarely-photographed vehicle. This Panzerbeobachtungswagen IV (PzBeobWg IV) is one of the five specialist artillery observation tanks delivered to SS-Panzerartillerie-Regiment 12 on 22 November 1944 and assigned to the regiment's I.Abteilung. The battalion took part in the Ardennes Offensive as part of Kampfgruppe Kuhlmann but I have been unable to discover the exact number of serviceable PzBeobWg IV involved or how many were lost. The official caption of this photograph states that it was taken near Weiswambach, a small town south of St.Vith, and the regiment's parent formation, 12.SS-Panzer-Division Hitler Jugend, was heavily engaged here in the early part of the assault. A reconstruction of this vehicle, as it may have looked when it was delivered to the regiment, is included in the Camouflage & Markings section of this book.

PZKPFW IV ALLOCATIONS, NOVEMBER 1944-APRIL 1945

The allocation of armoured vehicles was the responsibility of the various Heereszeugämter (HZA), or army ordnance offices, located throughout Germany, Austria and also at Danzig and Posen. Tanks were assigned by the HZA at the request of the Generalinspekteur der Panzertruppen and the HZA reported to the Heereswaffenamt (HWA), the army's ordnance directory, which oversaw all aspects of weapons development and supply. Sometimes, but rarely, vehicles were dispatched straight from the manufacturer as an expedient. Information relating to the allocation and transport of each shipment was recorded in Führungsliste, or supply lists, and provision was made on these lists to record the unit to which vehicles

Month	Unit	Quantity	Allocated by HZA	Shipped by HZA	Remarks
November 1944	2.SS-Panzer-Division	28	3.11.1944	14.11.1944	Delivered 8.11.1944. Originally planned 30.
	9.SS-Panzer-Division	28	3.11.1944	1.11.1944	Delivered 14.11.1944. Originally planned 30.
	2.Panzer-Division	14	4.11.1944	10.11.1944	Delivered 14.11.1944 (1).
	2.Panzer-Division	14	4.11.1944	12.11.1944	Delivered 16.11.1944 (1).
	10.SS-Panzer-Division	5	8.11.1944	20.11.1944	Panzerbeobachtungswagen. No delivery date.
	21.Panzer-Division	10	14.11.1944	16.11.1944	Noted that 29 sent previously.
	116.Panzer-Division	14	14.11.1944	18.11.1944	No delivery date recorded.
	Führer-Grenadier-Brigade	5	20.11.1944	30.11.1944	Panzerbeobachtungswagen. Delivered 5.12.1944.
	9.Panzer-Division	28	23.11.1944	26.11.1944	Delivered 28.11.1944.
	11.Panzer-Division	15	23.11.1944	25.11.1944	Delivered 30.11.1944.
	II.Abt., Panzer-Regiment 2	6	27.11.1944	29.11.1944	Delivered 2.12.1944.
	II.Abt., Panzer-Regiment Grossdeutchland (GD)	7	27.11.1944	30.11.1944	Delivered 4.12.1944.
	10.SS-Panzer-Division	34	28.11.1944	5.12.1944	Delivered 7.12.1944.
	Führer-Begleit-Brigade	3	30.11.1944	8.12.1944	Delivered 27.12.1944.
	21.Panzer-Division	5	30.11.1944	15.12.1944	Panzerbeobachtungswagen. No delivery date.
	9.Panzer-Division	4	30.11.1944	21.12.1944	Panzerbeobachtungswagen. No delivery date.
	116.Panzer-Division	4	30.11.1944	21.12.1944	Panzerbeobachtungswagen. No delivery date.
	Export to Hungary	10	30.11.1944	8.12.1944	Possibly diverted to 24.Panzer-Division.
	Conversion to Panzerbeobachtungswagen	20			Delivered direct to factory.

1. These two deliveries originated as a planned allocation of 30 vehicles.

Month	Unit	Quantity	Allocated by HZA	Shipped by HZA	Remarks
December 1944	1.Panzer-Division	2	4.12.1944	27.12.1944	Panzerbeobachtungswagen. Delivered 28.12.1944.
	23.Panzer-Division	2	4.12.1944	27.12.1944	Panzerbeobachtungswagen. Delivered 28.12.1944.
	24.Panzer-Division	2	4.12.1944	27.12.1944	Panzerbeobachtungswagen. Delivered 28.12.1944.
	13.Panzer-Division	1	4.12.1944	27.12.1944	Panzerbeobachtungswagen. Delivered 28.12.1944.
	11.Panzer-Division	17	5.12.1944	10.12.1944	Delivered 13.12.1944.
	11.Panzer-Division	2	5.12.1944	10.12.1944	Delivered 19.12.1944.
	116.Panzer-Division	5	5.12.1944	10.12.1944	Delivered 19.12.1944.
	Panzer-Lehr-Division	10	5.12.1944	10.12.1944	Delivered 27.12.1944.
	Panzer-Abteilung 208	31	5.12.1944	8.12.1944	No delivery date.
	Export to Hungary	10	7.12.1944	17.12.1944	Delivered 19.12.1944.
	7.Panzer-Division	17	8.12.1944	20.12.1944	Delivered 22.12.1944.
	Heeresgruppe Süd	7	10.12.1944	16.12.1944	Returned from repair. Delivered 17.12.1944.
	11.Panzer-Division	5	13.12.1944	16.12.1944	Delivered 26.12.1944.
	11.Panzer-Division	6	13.12.1944	20.12.1944	Delivered 27.12.1944.
	21.Panzer-Division	2	24.12.1944	25.12.1944	Delivered 27.12.1944.
	12.SS-Panzer-Division	2	31.12.1944	12.1.1944	Panzerbeobachtungswagen. No delivery date.
	Conversion to Panzerbeobachtungswagen	25			Delivered direct to factory.
	1.SS-Panzer-Division	11	31.12.1944	3.1.1945	Delivered 13.1.1945 (1).
	12.SS-Panzer-Division	5	31.12.1944	3.1.1945	Delivered 15.1.1945 (1).

1. Probably allocated very late in the day as they are both included in the January list.

Month	Unit	Quantity	Allocated by HZA	Shipped by HZA	Remarks
January 1945	Export to Hungary	12	2.1.1945	6.1.1945	Delivered 7.1.1945.
	Reserve Mayen (Pz-Stützpunkt Mayen)	17	2.1.1945	13.1.1945	Delivered 21.1.1945.
	Reserve Mitte (Pz-Stützpunkt Mayen)	17	4.1.1945	10.1.1945	Delivered 16.1.1945.
	24.Panzer-Division	31	6.1.1945	15.1.1945	Delivered 12.1.1945.
	II.Abt., Panzer-Regiment 10	14	6.1.1945	10.1.1945	Delivered 1.1.1945.
	10.SS-Panzer-Division	8	8.1.1945	15.1.1945	Delivered 18.1.1945.
	Nachschub Südwest	17	9.1.1945	28.1.1945	No delivery date recorded.
	12.SS-Panzer-Division	2	11.1.1945	2.2.1945	Panzerbeobachtungswagen. Delivered 19.2.1945 (1).
	II.Abt., Panzer-Regiment 9	14	15.1.1945	19.1.1945	No delivery date recorded.
	4.Panzer-Division	28	19.1.1945	24.1.1945	No delivery date recorded.
	19.Panzer-Division	3	21.1.1945	24.1.1945	Panzerbefehlswagen. No delivery date.
	4.Panzer-Division	3	24.1.1945	8.2.1945	Panzerbefehlswagen. No delivery date.
	8.Panzer-Division	17	26.1.1945	3.2.1945	No delivery date recorded.

1. Figures given for Panzerbeobachtungswagen allocations, particularly after January 1945, should be regarded as provisional as the differentiation between vehicles based on Pzkpfw IV and Pzkpfw III chassis is not always clear.

were assigned, the number of vehicles and the date the allocation was confirmed, the date transport was arranged and the date on which the shipment was actually dispatched. There were also columns for railway shipping numbers and to advise the date that the shipment arrived at its destination. Unfortunately, for our purposes, the latter is often left blank and the lists themselves should not be regarded as unconditionally accurate. Nevertheless, they are a valuable source of information and the chart below was compiled largely from the Pzkpfw IV Zuführungsliste. Note that this chart includes all allocations for the period in question including those made to units serving on the Eastern and Southern Fronts.

Month	Unit	Quantity	Allocated by HZA	Shipped by HZA	Remarks
February 1945	21.Panzer-Division	16	4.2.1945	7.2.1945	Delivered 10.2.1945.
	Panzer-Division Holstein	14	4.2.1945	2.2.1945	Delivered 7.2.1945 (1).
	Panzer-Division Holstein	3	4.2.1945	11.2.1945	Panzerbefehlswagen. Delivered 16.2.1945 (2).
	Panzer-Division Jüterbog	See Panzerjäger-Abteilung 510 below.			
	Panzer-Division Schlesien	See Panzer-Abteilung 303 below.			
	Sturmgeschütz-Brigade 190	3	4.2.1945	7.3.1945	Delivered 22.3.1945.
	Sturmgeschütz-Brigade 210	4	4.2.1945	7.3.1945	Delivered 14.3.1945 (3).
	Sturmgeschütz-Brigade 276	3	4.2.1945	7.3.1945	Delivered 22.3.1945.
	Sturmgeschütz-Brigade 300	4	4.2.1945	7.3.1945	Delivered 13.3.1945.
	Sturmgeschütz-Brigade 311	4	4.2.1945	7.3.1945	Delivered 13.3.1945.
	Sturmgeschütz-Brigade GD	31	4.2.1945	9.2.1945	Delivered 12.2.1945.
	17.Panzer-Division	16	7.2.1945	9.2.1945	Delivered 11.2.1945
	1.SS-Panzer-Division	3	8.2.1945	15.2.1945	Delivered 16.2.1945.
	2.SS-Panzer-Division	11	8.2.1945	13.2.1945	Delivered 14.2.1945.
	10.SS-Panzer-Division	9	8.2.1945	9.2.1945	Delivered 22.2.1945.
	12.SS-Panzer-Division	9	8.2.1945	15.2.1945	Delivered 16.2.1945.
	20.Panzer-Division	10	12.2.1945	15.2.1945	Delivered 18.2.1945.
	8.Panzer-Division	10	12.2.1945	15.2.1945	Delivered 25.2.1945.
	25.Panzer-Division	21	15.2.1945	18.2.1945	Delivered 23.2.1945.
	Panzer-Abteilung 303	20	16.2.1945	21.2.1945	Delivered 24.2.1945 (4).
	1.SS-Panzer-Division	2	16.2.1945	6.3.1945	Panzerbeobachtungswagen. No delivery date.
	2.SS-Panzer-Division	2	16.2.1945	6.3.1945	Panzerbeobachtungswagen. No delivery date.
	9.SS-Panzer-Division	2	16.2.1945	6.3.1945	Panzerbeobachtungswagen. No delivery date.
	12.SS-Panzer-Division	2	16.2.1945	6.3.1945	Panzerbeobachtungswagen. No delivery date.
	19.Panzer-Division	20	19.2.1945	2.3.1945	Delivered 4.3.1945.
	Panzer-Abteilung 303	1	20.2.1945	21.2.1945	Delivered 24.2.1945.
	Panzer-Abteilung 303	10	22.2.1945	2.3.1945	No delivery date recorded (5).
	Panzerjäger-Abteilung 510	1	20.2.1945	2.3.1945	No delivery date recorded (6).
	Panzer-Division Hermann Göring	20	23.2.1945	5.3.1945	Delivered 7.3.1945.
	Conversion to Panzerbeobachtungswagen	25			Delivered direct to factory.

1. Transported directly to the division from Stettin. 2. Delivered from the Linz arsenal which was located 5 kilometres north-west of the Nibelungenwerk at Sankt Valentin. The division also took over twenty-nine Pzkpfw IV tanks from 233.Reserve-Panzer-Division. 3. The battalion's strength report of 8 April 1945 does not mention any Pzkpfw IV tanks. 4. Attached to Panzer-Division Schlesien and renamed Panzer-Abteilung Schlesien on 22 February 1945. 5. At least one authoritative source suggests that these vehicles were in fact shipped on 2.3.1945 after the battalion had been renamed. 6. Used to form Panzerjäger-Abteilung Glienecke on 17 February 1945 and renamed Panzer-Abteilung Jüterbog three days later.

Month	Unit	Quantity	Allocated by HZA	Shipped by HZA	Remarks
March 1945	25.Panzer-Division	10	2.3.1945	6.3.1945	No delivery date recorded.
	20.Panzergrenadier-Division	20	5.3.1945	10.3.1945	Delivered 14.3.1945.
	Panzer-Division Feldherrnhalle	5	7.3.1945	12.3.1945	Delivered 14.3.1945 (1).
	13.Panzer-Division	4	7.3.1945	17.3.1945	Panzerbeobachtungswagen. No delivery date (1).
	13.Panzer-Division	20	9.3.1945	12.3.1945	Delivered 14.3.1945 (1).
	11.Panzer-Division	12	16.3.1945	17.3.1945	Allocated through Nachschub OB West (2).
	116.Panzer-Division	20	20.3.1945	26.3.1945	Allocated through Nachschub OB West (3).
	9.Panzer-Division	20	25.3.1945	29.3.1945	Allocated through Nachschub OB West (4).
	17.Panzer-Division	3	30.3.1945		Panzerbeobachtungswagen. No delivery date.
	19.Panzer-Division	3	30.3.1945		Panzerbeobachtungswagen. No delivery date.
	20.Panzer-Division	4	30.3.1945		Panzerbeobachtungswagen. No delivery date.

1. The allocations were made to these two divisions as part of the creation of Panzerkorps Feldherrnhalle. 2. No delivery date recorded. Nachschub OB West was the supply hub for the Western Front. 3. A handwritten notation states that these were delivered on 3.4.1945. 4. No delivery dated recorded.

Month	Unit	Quantity	Allocated by HZA	Shipped by HZA	Remarks
April 1945	9.Panzer-Division	6	9.4.1945	9.4.1945	Returned from repair.
	7.Panzer-Division	13	14.4.1945	18.4.1945	Originally planned 20. No delivery date recorded (1).
	Heeresgruppe Süd	4	11.4.1945	11.4.1945	To 13.Panzer-Division.
	Heeresgruppe Süd	4	11.4.1945	11.4.1945	To Panzer-Division Feldherrnhalle.
	Heeresgruppe Süd	10			No details other than planned allocation recorded (2).

1. No allocation date is recorded on the Führungsliste and 14 April 1945 was actually the date on which Generalinspekteur der Panzertruppen requested that the entire Nibelungenwerk production run be allocated to the division's Panzer-Regiment 25. Although the exact delivery date is not known, the division reported that thirteen vehicles were on hand by 19 April 1945. 2. All entries for April 1945 were made by hand and are in some cases unclear. It is likely that these ten vehicles were never shipped as the total given in the Führungsliste for the month is twenty-seven.

Although these images are sometimes associated with the Allied air raid of 23 March 1945 it is more likely they depict the damage done to the Nibelungenwerk at Sankt Valentin on 16 October 1944. All the vehicles shown on this page are Pzkpfw IV ausf J models and all are finished in the factory-applied camouflage scheme that was introduced in August or September 1944. Note that the pattern is extended to the wire mesh screens of the Drathgeflechtschürzen. The three hangers on the left side of the hull rear plate, which are clearly visible in the photograph below, were an identifying feature of Pzkpfw IV tanks produced at this plant.

1

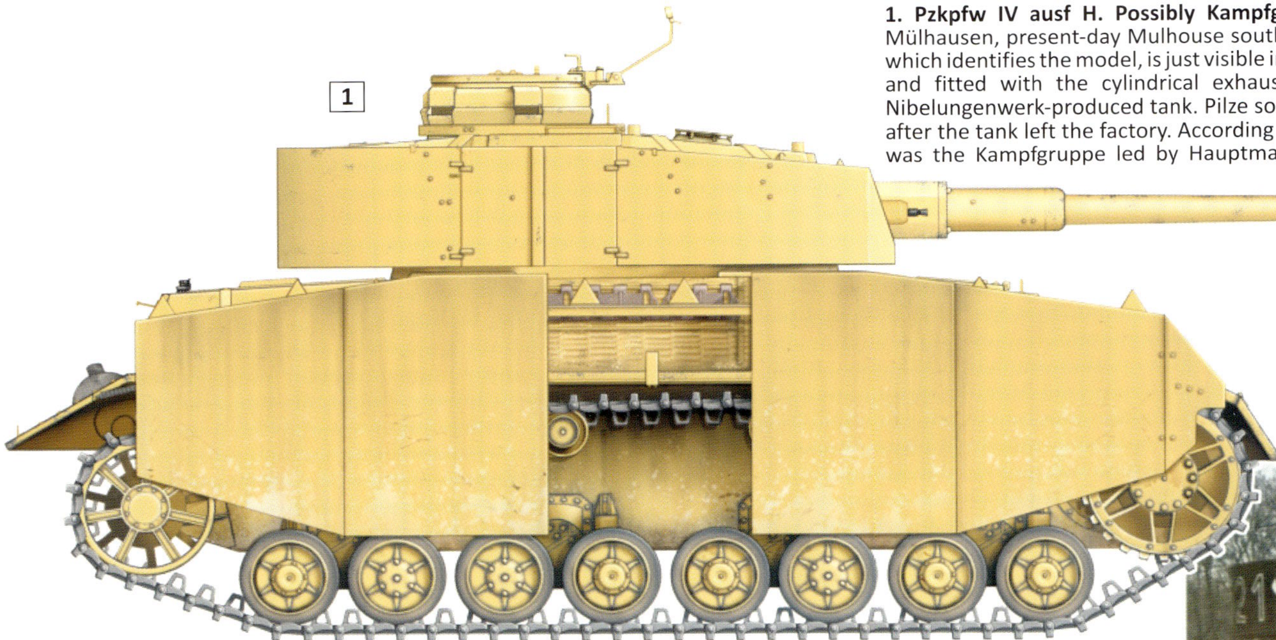

1. Pzkpfw IV ausf H. Possibly Kampfgruppe te Heesen. Eastern France, late 1944. Photographed outside Mülhausen, present-day Mulhouse south of Colmar, the exhaust muffler for the electrical turret drive generator, which identifies the model, is just visible in the photograph on which our illustration is based. It is coated in Zimmerit and fitted with the cylindrical exhaust muffler and the Series II Schürzen, with the cut-out denoting a Nibelungenwerk-produced tank. Pilze sockets have been welded to the turret roof, almost certainly at some time after the tank left the factory. According to the surviving OKW situation maps, the only armoured unit in the area was the Kampfgruppe led by Hauptmann Paul te Heesen, the commander of Panzer-Abteilung 2106, which is said to have contained a number of Pzkpfw IV and Pzkpfw III tanks, although their origin is something of a mystery as the battalion was equipped throughout with Panthers and they are not mentioned again after the battles of December 1944. This vehicle was in good condition when it was photographed during the spring of 1945 and all the plates of the hull Schürzen were in place although I have omitted one to show the Zimmerit and spare tracks.

3

2. Pzkpfw IV ausf J. Unit uncertain. Eastern France, late 1944. Although the style of turret number is similar to that used by Panzer-Regiment 33 (see page 23), the Pzkpfw IV tanks of that regiment were all assigned to I.Abteilung. Another possible candidate is II.Abteilung, Panzer-Lehr-Regiment 130 (3), but the Pzkpfw IV tanks of that unit had been concentrated in the battalion's 6.Kompanie and 8.Kompanie during the retreat through France after the Normandy battles. There is also the remnants of a large number 3, which is faded and partly covered by the roughly applied Rotbraun bands of camouflage, but this does nothing to help with any identification. Note the patches of Zimmerit and that all the brackets for the hull Schürzen have been removed.

2

1. Pzkpfw IV ausf G. 1.Kompanie, Panzer-Regiment 55. Denmark, late 1944. The regiment's parent formation, 233.Reserve-Panzer-Division, was stationed in Denmark and by late 1943 had been allocated twenty-two Pzkpfw IV tanks. In early 1945 all but five of the tanks were returned to Germany and used to create the short-lived Panzer-Division Holstein and Panzer-Division Clausewitz.

1

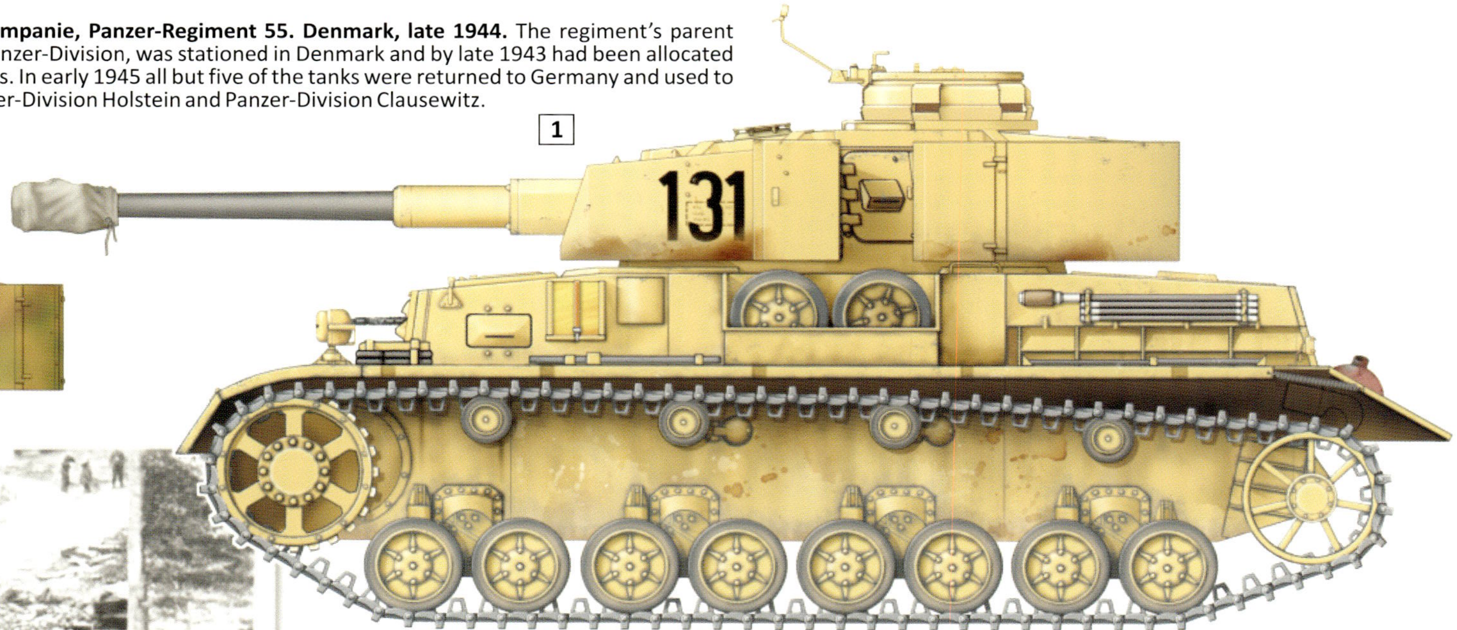

4

3

Above: In Denmark none of the tanks of Panzer-Regiment 55 were camouflaged and the dark colour along the lower edge of the turret Schürzen is a stain of some kind. Gun barrels were coated in a heat-resistant lacquer of a very dark, almost black, colour and these were commonly left unpainted. Note that the company number has been applied over the loading label.

2

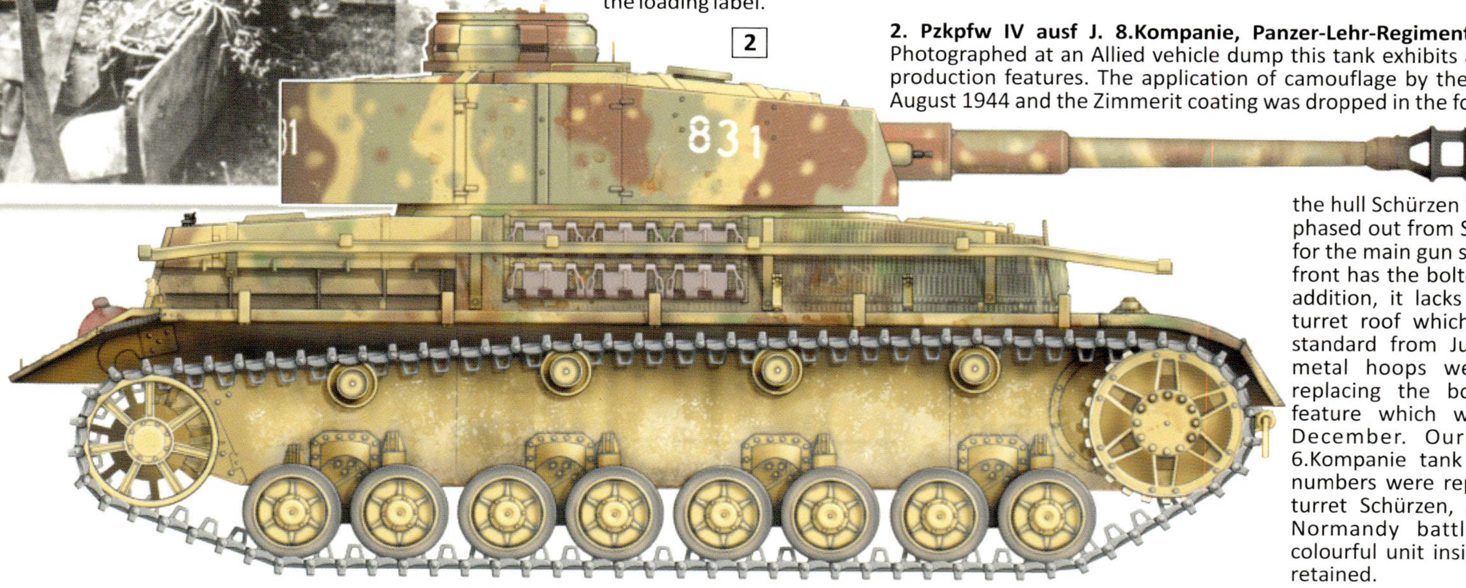

2. Pzkpfw IV ausf J. 8.Kompanie, Panzer-Lehr-Regiment 130. France, late 1944. Photographed at an Allied vehicle dump this tank exhibits a mixture of early and late production features. The application of camouflage by the assembly plants began in August 1944 and the Zimmerit coating was dropped in the following month. The rail for the hull Schürzen is of the early Series I type, phased out from September 1943, the cover for the main gun sight is in place and the hull front has the bolted-on towing couplings. In addition, it lacks the Pilze sockets on the turret roof which, officially at least, were standard from June 1944 but has simple metal hoops welded to the hull front replacing the bolted-on lifting hooks, a feature which was not introduced until December. Our photograph (3) of a 6.Kompanie tank suggests that the turret numbers were repeated on the rear of the turret Schürzen, as they were during the Normandy battles, but the battalion's colourful unit insignia (4) was probably not retained.

1

2

1. Pzkpfw IV ausf J. 7.Kompanie, SS-Panzer-Regiment 1. Belgium, December 1944. The regiment was largely rebuilt after the battles in Normandy and the subsequent retreat and most of the II.Abteilung tanks seem to have been painted in one of the factory-applied camouflage schemes. This tank was famously filmed as it passed a column of captured US Army soldiers and is almost certainly the vehicle which was later abandoned and photographed in early 1945 (2). Note the Drahtgeflechtsschürzen, made up of wire mesh screens, often referred to today as Thoma Schürzen or Thoma Schild. These are examined in detail on page 58.

3

4

3. Pzkpfw IV ausf J. 6.Kompanie, SS-Panzer-Regiment 1. Belgium, December 1944. Although the Drahtgeflechtsschürzen were introduced from September 1944 the earlier Series II version, shown here, could be seen until the end of the war. The factory-applied camouflage scheme depicted here and above are highly speculative on my part as the photographs used to create both illustrations contain restricted views of both vehicles. The rear of this 6.Kompanie tank was, however, quite clear and also showed that the company numbers were of the same style and depicted in the same manner as they had been in Normandy (4).

3
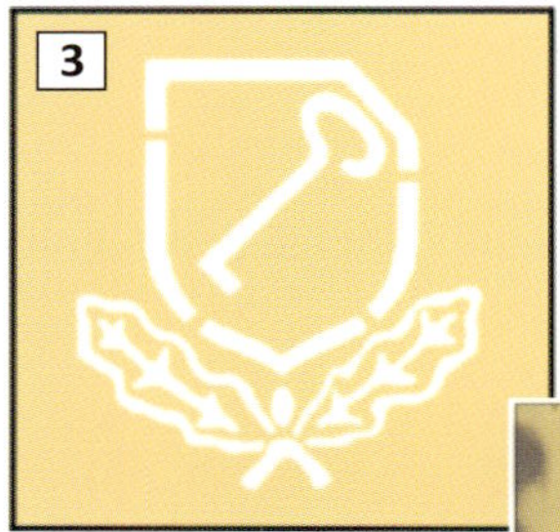

1

2

5

4

1. Pzkpfw IV ausf J. 7.Kompanie, SS-Panzer-Regiment 1. Belgium, December 1944. Although this tank is badly damaged in the photograph used to create this illustration (2) an interesting variation on the factory-applied camouflage can be discerned and our image also shows that the centre of the Balkenkreuz, which appears to be light or missing in other photographs, is indeed black. The wheels and lower hull were painted on some tanks but the rendering here is largely speculative. The unit insignia of 1.SS-Panzer-Division (3) was probably not carried by the battalion's tanks and I have included it here as a matter of interest only.

4. Panzerbefehlswagen IV ausf J. Stab, II.Abteilung, SS-Panzer-Regiment 1. Belgium, December 1944. This vehicle, Fgst Nr.90200, was probably one of seventeen Pzkpfw IV tanks converted to command vehicles in August and September 1944. Note the TSR.1 periscope, here in the raised position, in front of the commander's cupola and the Sternantenna D for the FuG 8 radio on the hull rear. The Stützkorb für Antennenfuss, shown in the image at right (5) was intended to protect the antenna insulator and is seldom seen in contemporary photographs. Note the non-standard Balkenkreuz on the turret Schürzen sides and rear. This vehicle is today on display at the Aberdeen Proving Ground Museum, Maryland.

2

1

1. Pzkpfw IV ausf J. 5.Kompanie, Panzer-Regiment 16. Belgium, December 1944. This tank was disabled in the fierce fighting for the town of Hotton, about 35 kilometres south of Liège, in the days leading up to Christmas 1944. Although badly burnt in places, the camouflage scheme is reasonably clear and it appears to have been extended to the lower hull behind the roadwheels, which are also shown painted in the colours of Olivgrün, Rotbraun and Dunkelgelb. The unit insignia of 116.Panzer-Division, the regiment's parent formation, is shown here (2) although it is doubtful that any of the Pzkpfw IV tanks of II.Abteilung carried it.

4

3

3. Pzkpfw IV ausf J. Panzer-Abteilung 115. Disabled and abandoned in the village of Champs, 4 kilometres north of Bastogne, this tank is one of the vehicles attached to Kampfgruppe Maucke for the Ardennes Offensive. Note the rack to hold track links on the hull side at the front and that other links have been fitted across the hull front. These can be seen clearly in the photograph on page 10. Note also that one of the return rollers is missing. The name Lustmolch (4) can be roughly translated as lecher. In the photograph on which this illustration is based what may be a number 5 is just visible behind the Balkenkreuz on the turret Schürzen and a single kill ring on the barrel.

1. Panzerbeobachtungswagen IV ausf J. SS-Panzer-Artillerie-12. Luxembourg, early 1945. The artillery observation version of the Pzkpfw IV is described in the Technical Details & Modifications section of this book. They were fitted with the TSR.1 periscope, as were the Panzerbefehlswagen command tanks, but the cupola was replaced by the model used with the Sturmgeschütz III assault gun (2). The artillery regiment of 12.SS-Panzer-Division was allocated five of these specialist vehicles in November 1944 and this example was abandoned near Weiswambach, south of St. Vith. The camouflage scheme of broad bands of colour was almost certainly applied at the factory but the darker shades are far more prominent than other examples such as those depicted on page 23 and page 26.
This vehicle is also shown in the photograph on page 13.

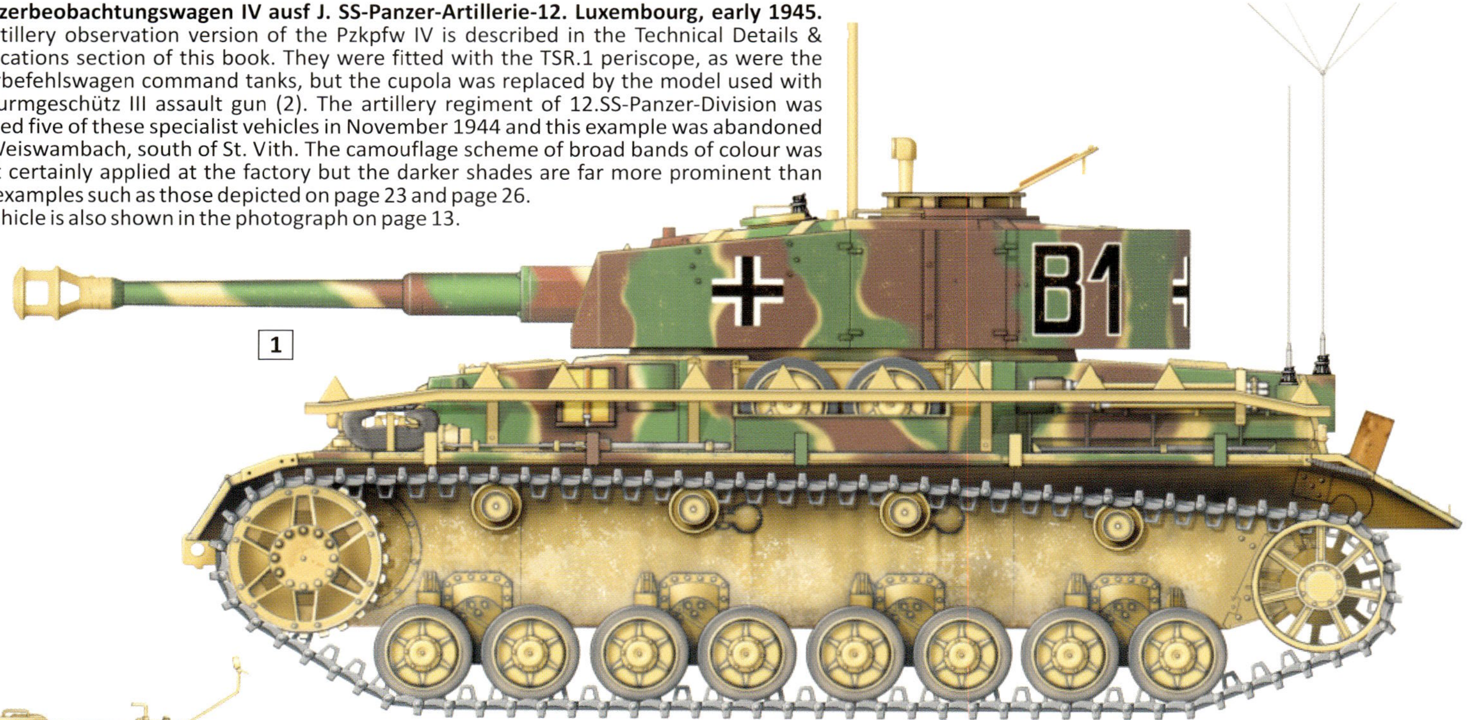

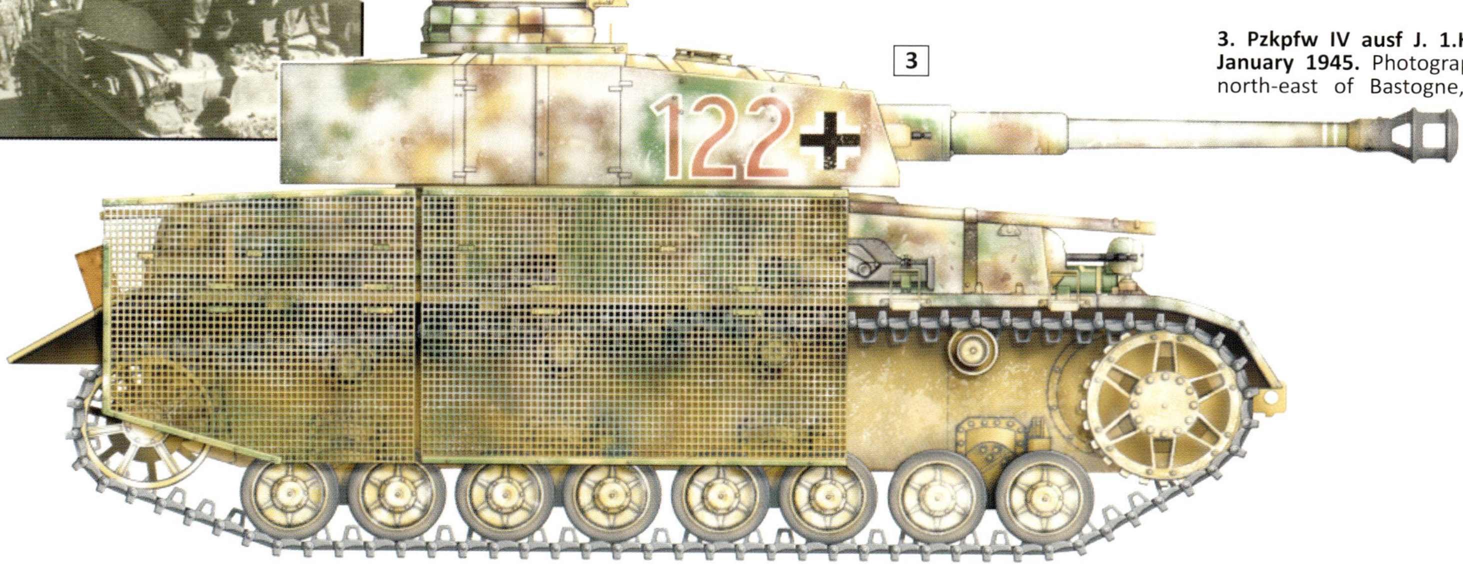

3. Pzkpfw IV ausf J. 1.Kompanie, Panzer-Abteilung 115. Belgium, January 1945. Photographed near Houffalize, about 15 kilometres north-east of Bastogne, this tank was badly damaged but the photograph on which this illustration is based clearly shows the turret number and parts of the Drahtgeflechtsschürzen rail allowing a fairly accurate reconstruction to be made. The kill rings on the barrel of the 7.5cm gun are in the same style as the tank of Panzer-Abteilung 115 depicted on the previous page and the position of the company number was common within the battalion. The latter is often rendered as a white outline only but close examination of the original photograph (4) suggests that the centres were a dark colour as they certainly were when the battalion arrived in France (5) in late 1943.

1

2

5

1. Pzkpfw IV ausf J. 3.Kompanie, Panzer-Regiment 33. Belgium, early 1945. This tank may be one of the vehicles lost when the regiment's I.Abteilung was engaged in the fighting for Langlire, a village between Bastogne and Malmedy. It is fitted with the swivelling hatch for the commander's cupola and Pilze sockets but has the earlier fume extractor cover. The company number is rendered here as yellow but could be a very faded and dirty white. The attractive unit insignia, show enlarged at right (2), commemorated the honour title of Panzer-Regiment Prinz Eugen granted in March 1943 to acknowledge the regiment's Austrian origins.

4

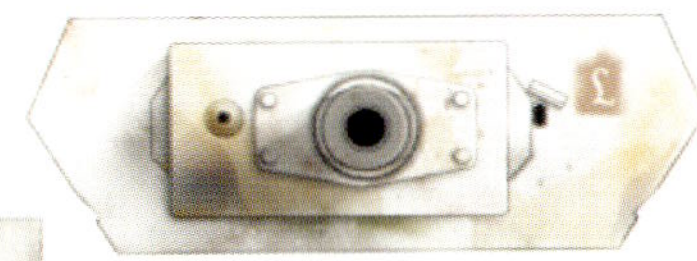

3

3. Pzkpfw IV ausf J. Unit uncertain but possibly 6.Kompanie, Panzer-Lehr-Regiment 130. Germany, early 1945. Abandoned after the fighting for the Remagen bridgehead, this tank has a number of late production features including the extended hull, Flammentöter exhaust mufflers, three return rollers, swivelling hatch for the commander's cupola and bracket for the Orterkompass. Patches of whitewash camouflage are clearly visible in our photograph but I have exaggerated its presence here slightly to show how this vehicle would have appeared when it was operational. It has been suggested that the small rectangle of dark colour next to the sight aperture for the main gun (4), just visible in the photograph used to create this illustration, may show the unit insignia of Panzer-Regiment 15 of 11.Panzer-Division. But I feel that a more likely candidate may be Panzer-Lehr-Regiment 130 which was equipped with Pzkpfw IV tanks and was also in this area although the image is too faint for any definitive identification to be made. The unit insignia of Panzer-Lehr-Division, the regiment's parent formation, is shown at right (5).

1. Panzerbefehlswagen IV ausf G. Unit uncertain. Germany, early 1945. The identification of this vehicle as a command tank is based on the presence of the blanked-off machine-gun aperture in the gun mantlet, which can be seen in our photograph (2), what appears to be an antenna insulator, just visible between to two men at left, and the company number. Eighty-eight tanks returned to Germany for repairs between March and July 1944 were converted to Panzerbefehlswagen and this tank may be one of those. It could also be a vehicle modified by a unit workshop as examples with non-standard aerial arrangements can be seen in contemporary photographs.

1

3

2

The caption of the US Army Signal Corps photograph on which our illustration is based (2) states only that it was being examined by men of the US 104th Infantry Division. A comparison of the surviving OKW and US Army situation maps shows that Kamppfgruppen of Panzer-Regiment 15 and Panzer-Regiment 33 were operating in the same area in late March when this photograph may have been taken. The name Rita (3) probably refers to a girlfriend or sister.

3

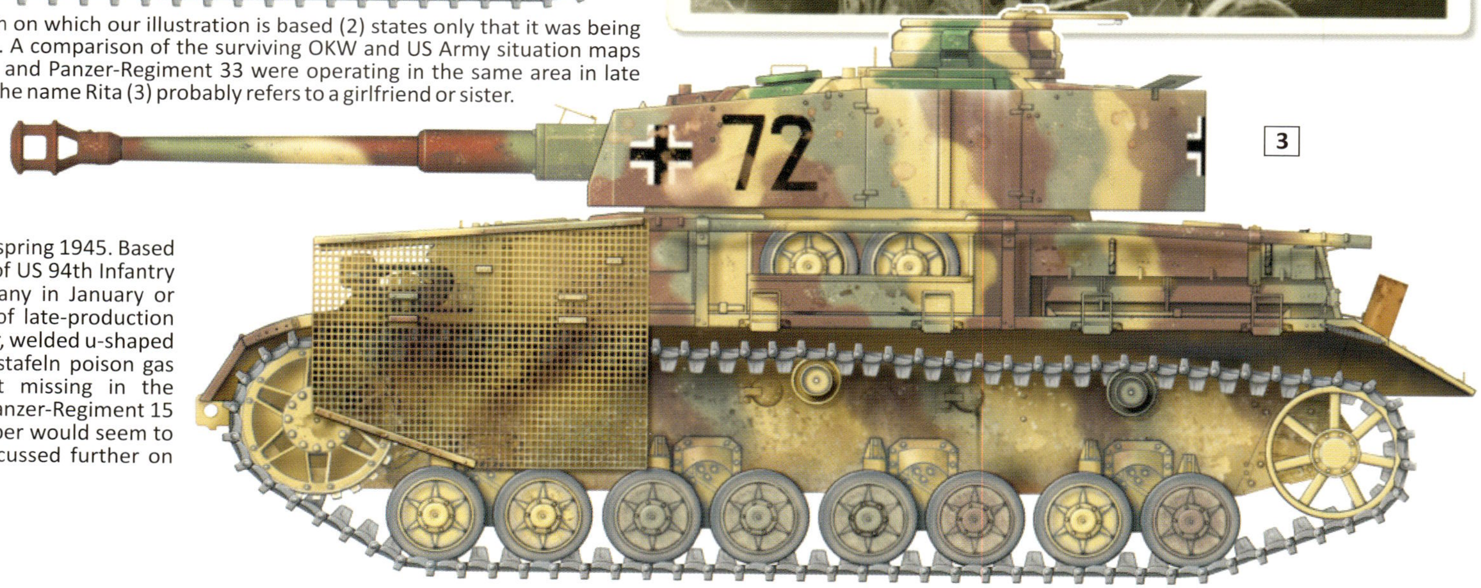

3. Pzkpfw IV ausf J. Panzer-Regiment 15. Germany, spring 1945. Based on a snapshot said to have been taken by a soldier of US 94th Infantry Division in the Hunsrück region of western Germany in January or February 1945, this tank is fitted with a number of late-production features including three return rollers in place of four, welded u-shaped brackets on the hull front and the Lost-Erkennungstafeln poison gas detection panels, which are for the most part missing in the photograph. The division was opposed by tanks of Panzer-Regiment 15 until late February and the two-digit company number would seem to confirm this identification. These numbers are discussed further on page 4 and page 5.

2

3

4

1. Pzkpfw IV ausf J. Unidentified unit. Germany, spring 1945. This tank features a number of the late production modifications including the poison gas detection panels (2 and 3). Note the bracket for the Orterkompass fitted to the armoured cover of the fume extractor. It is unlikely that any of the compasses were issued but I have included an image here (4) of the 1/35 scale version offered by ROCHM Model.

1

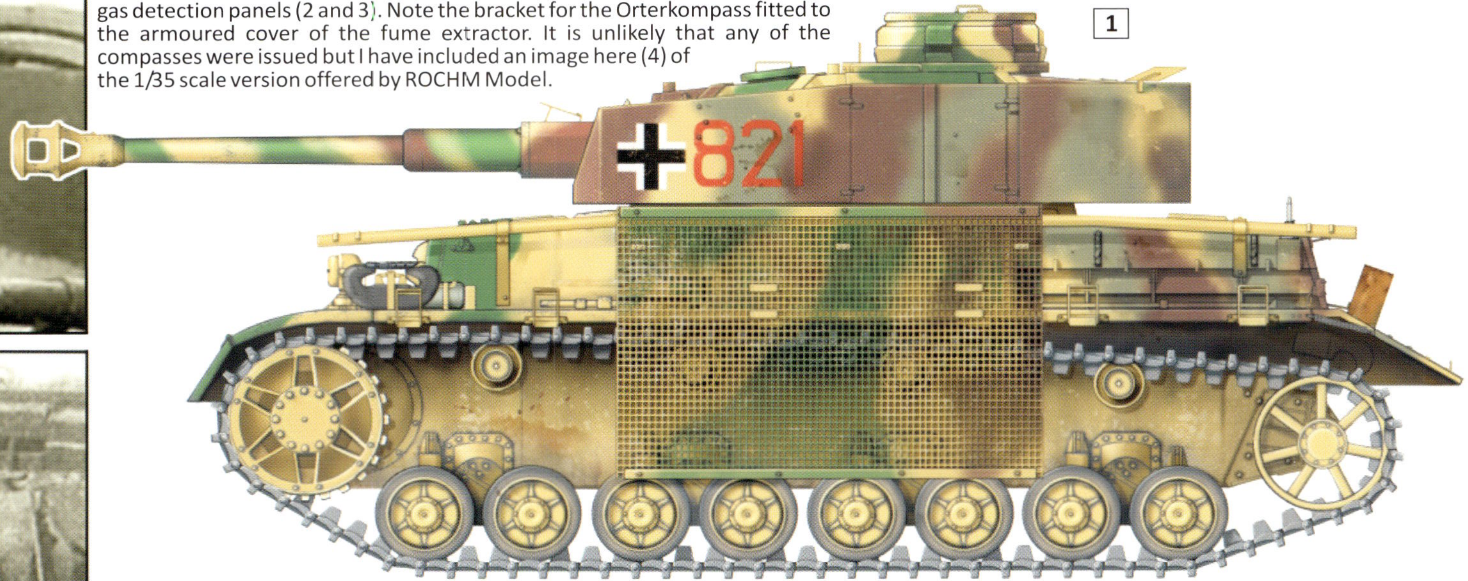

5. Pzkpfw IV ausf G. Panzer-Ersatz und Ausbildungs-Abteilung 35. Germany, spring 1945. Typical of the elderly vehicles pressed into service during the last weeks of the war, this tank was photographed in Bamberg where the battalion's workshop was located. Note that many fittings, including the jack block and gun cleaning tools, are missing.

5

1

1. Pzkpfw IV ausf J. Panzer-Regiment 15. Germany, spring 1945. This vehicle probably represents the final production model of the Pzkpfw IV tank. It is fitted with the swivelling hatch for the commander's cupola, the armoured cover for the fume extractor with Orterkompass bracket, Pilze sockets, metal loops in place of the hull lifting hooks and, not visible here, the brake access hatches without air intakes. The hull sides have been extended to form towing eyes, a modification that replaced the bolted-on towing couplings, and the factory-applied camouflage is a much-simplified version with hard edges and without contrasting dots of colour.

2

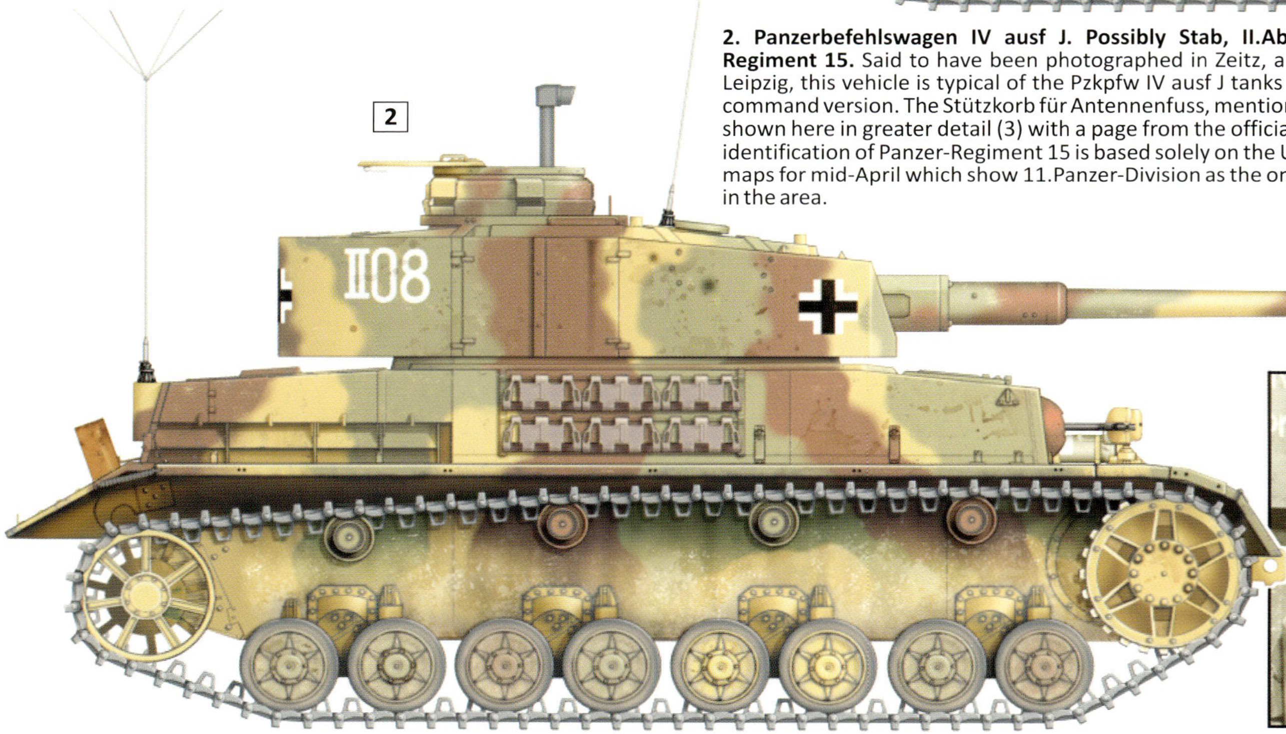

2. Panzerbefehlswagen IV ausf J. Possibly Stab, II.Abteilung, Panzer-Regiment 15. Said to have been photographed in Zeitz, a small town near Leipzig, this vehicle is typical of the Pzkpfw IV ausf J tanks converted to the command version. The Stützkorb für Antennenfuss, mentioned on page 20, is shown here in greater detail (3) with a page from the official manual (4). The identification of Panzer-Regiment 15 is based solely on the US Army situation maps for mid-April which show 11.Panzer-Division as the only armoured unit in the area.

4

3

1

2

1. Pzkpfw IV ausf J. Panzer-Regiment 15. Germany, May 1945. Photographed near the small town of Bad Kötzting, south of Regensburg, as the last remnants of this unit passed into captivity, this tank is a very late production model and displays the same modifications evident on the vehicle of the same regiment shown on the previous page and numbered 33. Note that the factory-applied camouflage has been painted with a somewhat softer edge in parts. A film made of the same event shows this tank towing a Pzkpfw IV ausf J finished in the same camouflage pattern and numbered 21 in the same, very rough, style. Although I have included many of the standard tools in the illustration thsy are missing in our photograph (2).

3

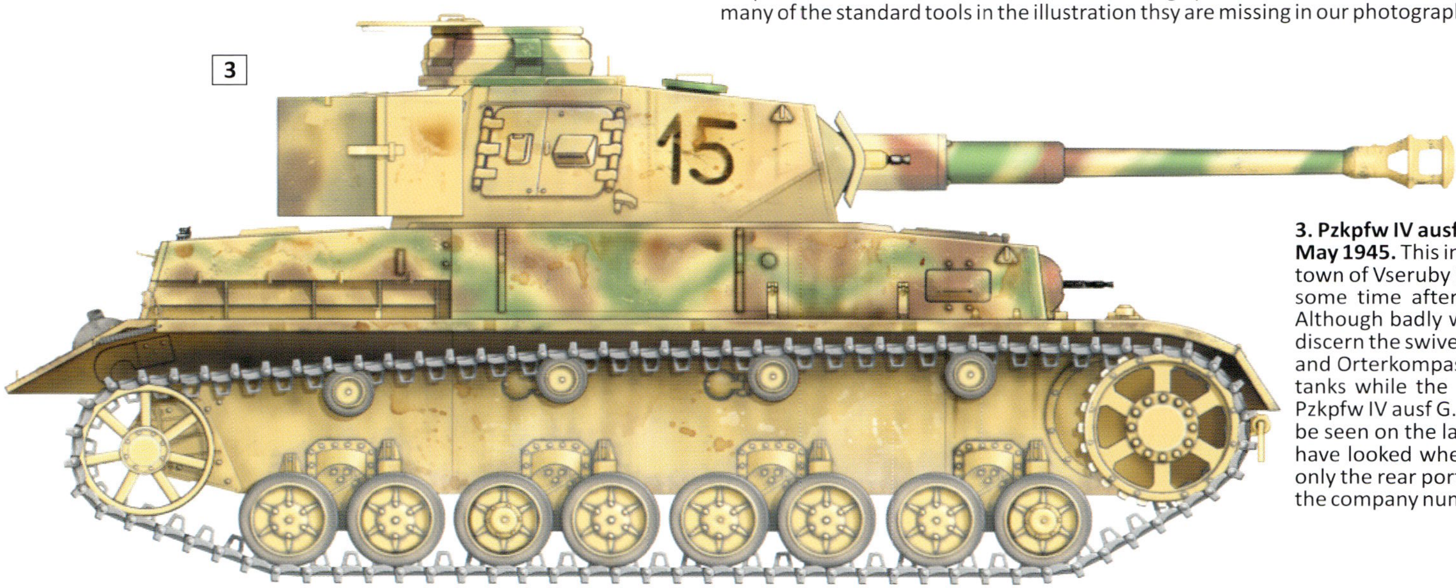

3. Pzkpfw IV ausf G/ausf J hybrid. Panzer-Regiment 15. Czechoslovakia, May 1945. This interesting vehicle was disabled and abandoned near the town of Vseruby in the present-day Czech Republic and photographed at some time after the end of the war, possibly the summer of 1946. Although badly worn, and many parts had been stolen, it is possible to discern the swivelling commander's hatch, armoured extractor fan cover and Orterkompass bracket common to late-production Pzkpfw IV ausf J tanks while the hull and chassis have been taken from a refurbished Pzkpfw IV ausf G. Faint traces of a distinct camouflage pattern can be still be seen on the latter and I have endeavoured to reconstruct how it may have looked when the tank was operational. It must have served with only the rear portion of the turret Schürzen otherwise the application of the company number would have been completely superfluous.

1

1. Pzkpfw IV ausf J. Unit uncertain. Eastern Germany, spring 1945. Said to have been photographed in the town of Glashütte, south of Dresden, this vehicle was produced at some time after January 1945 when the Lost-Erkennungsstafeln poison gas detection cards were introduced into the assembly process. German author Wolfgang Fleischer claims that this tank is from SS-Panzer-Regiment 10, elements of which retreated from Spremberg, reaching Morwitz on 3 May 1945. But most sources agree that the regiment's 1.Kompanie was equipped with Panthers and in addition Morwitz is almost 70 kilometres to the east of Glashütte. The surviving US Army situation maps, while not always accurate, indicate that parts of Panzer-Regiment 22 and Panzer-Regiment 102 were much closer and both would be possible candidates.

2. Pzkpfw IV ausf J. 1.Kompanie, Panzer-Regiment 33. Our illustration is based on a photograph (3) which is said to depict the tank commanded by Leutnant Ludwig Bauer who led 1.Kompanie until mid-April 1945. Note the poison gas detection panels and bracket for the Orterkompass fixed to the armoured cover of the fume extractor. The regiment's Prinz Eugen formation badge is just visible in our photograph as are the air intakes of the early brake access hatches which were dropped from production in February 1945. This tank is probably one of the vehicles sent to the regiment in late March and early April 1945.

2

3

Above: Leutnant Ludwig Bauer, at right, with other officers of I.Abteilung photographed before 17 April 1945 when the last of the battalion's tanks were destroyed in the Ruhr Pocket. Bauer served with Panzer-Regiment 33 from 1941 and is claimed to be, somewhat controversially, the last soldier of 9.Panzer-Division, the regiment's parent formation, to receive the Ritterkreuz for an action in which he took part on 29 April 1945. The officer on Bauer's left may be Hauptmann Simon, the battalion commander.

PANZER IV

AUSF J LATE PRODUCTION

ROY WELLS

1/35 SCALE

Roy's model is based on the Tamiya 1/35 scale kit, shown above, with figures from Alpine Miniatures. Humbrol enamels were used to paint the intricate base, both figures and the tank. The single remaining Schürzen plate was coated with patches of hair spray and painted with a white acrylic. When dry, the hairspray was removed leaving the weathered appearance. The images at the bottom of the page show parts of the groundwork and amply illustrate the attention to detail which characterises Roy's work.

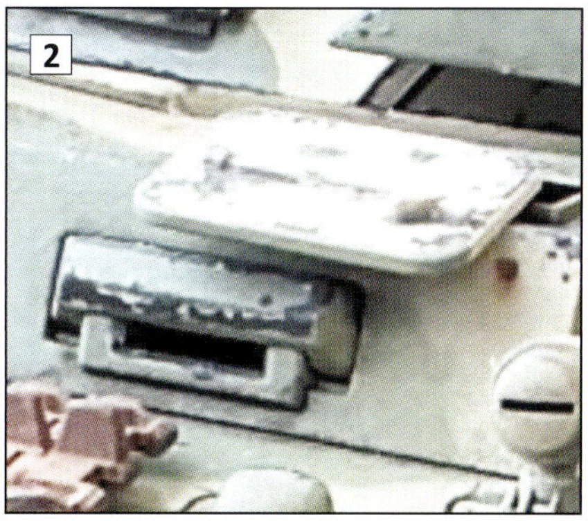

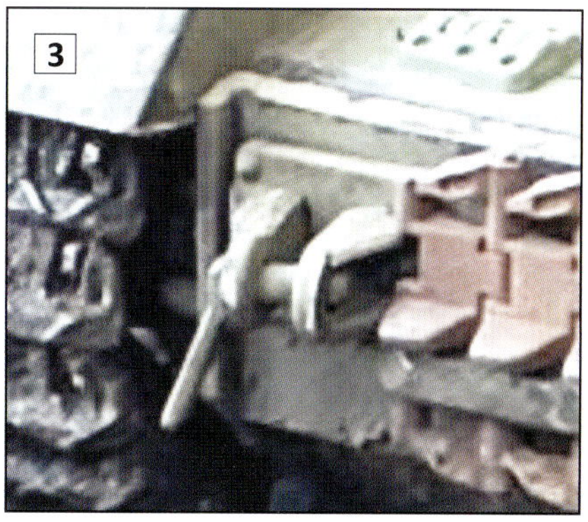

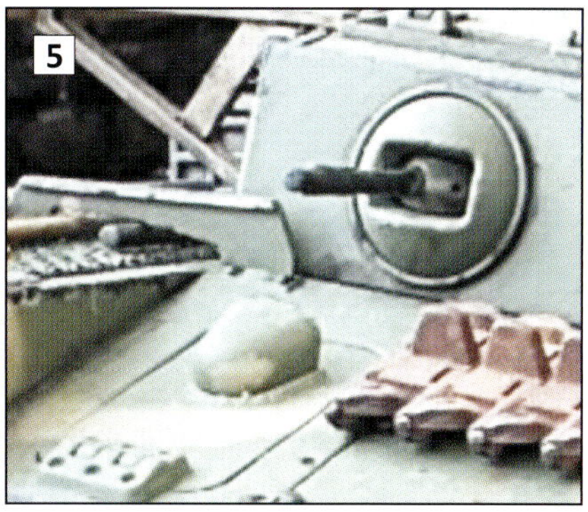

Above: Details of Roy's completed Pzkpfw IV model including, 1. The commander's cupola and turret stowage bin. 2. The opened driver's hatch and visor. 3. Bolted-on towing coupling. 4. Cylindrical exhaust muffler and filler cap. 5. Kugelblende machine-gun mount. 6. The vehicle's roadwheels and subtle mud coating on the lower hull. Below: The completed model on its base showing the realistic water effect and debris.

Another view of the completed diorama. Roy built the partly destroyed bridge from pieces of a Verlinden product marketed as a harbour wall and the stonework were created from a sheet of polythene foam. The remains of the road was made from moulded dental plaster onto which Roy carved a distressed brick effect. The area around the tank was built up from plaster and textured paste from the extensive MIG range, sprinkled with pebbles and fine stones. The water effect was created using Envirotex Lite, a two-part epoxy resin.

PANZER IV

JULY 1944 PRODUCTION

KUNIHITO KAWAMURA

1/35 SCALE

Kunihito's model is based on the Cyberhobby 1/35 scale kit, shown at left. The turret Schürzen and brackets are from Voyager Model while the Series II Schürzen rail and most of the photo-etched details fitted to the hull and turret are from Griffon Model. The tracks are individual metal links produced by Friulmodel. The realistic Zimmerit texture was achieved by adding thin layers of Tamiya putty and working it with a homemade roller.

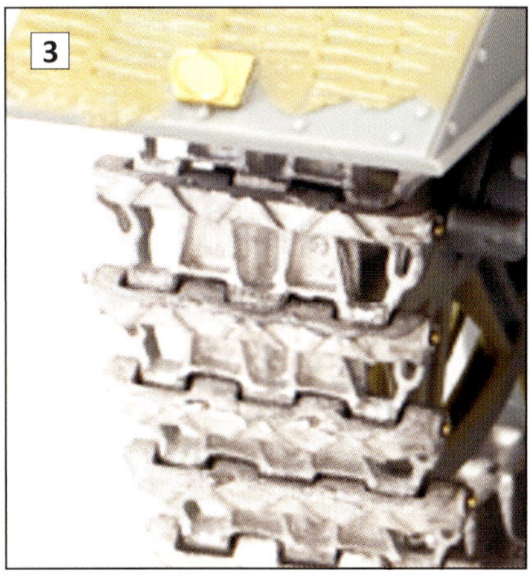

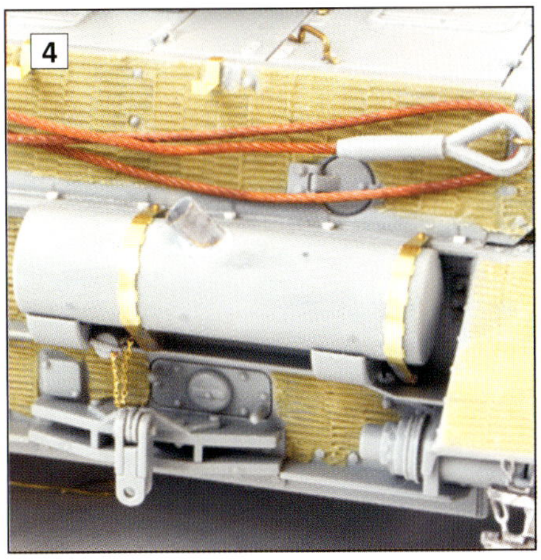

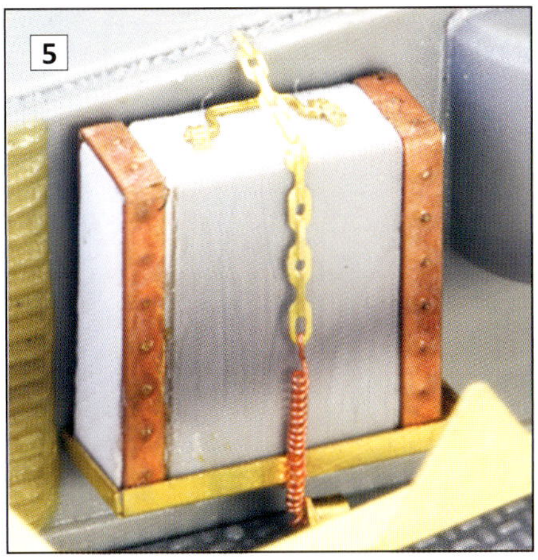

Above: Details of the fully built model. 1. The Series II Schürzen rails and turret Schürzen with opened access doors. 2. The commander's cupola with added scratch-built details. 3. The Friulmodel metal tracks. 4. The hull rear with towing cable from Eureka XXL Model Accessories and Griffon Model photo-etched parts. 5. The jack block and its retaining bracket were scratch-built by Kunihito. 6. The hull front and right side fender showing the pattern of Zimmerit texture.

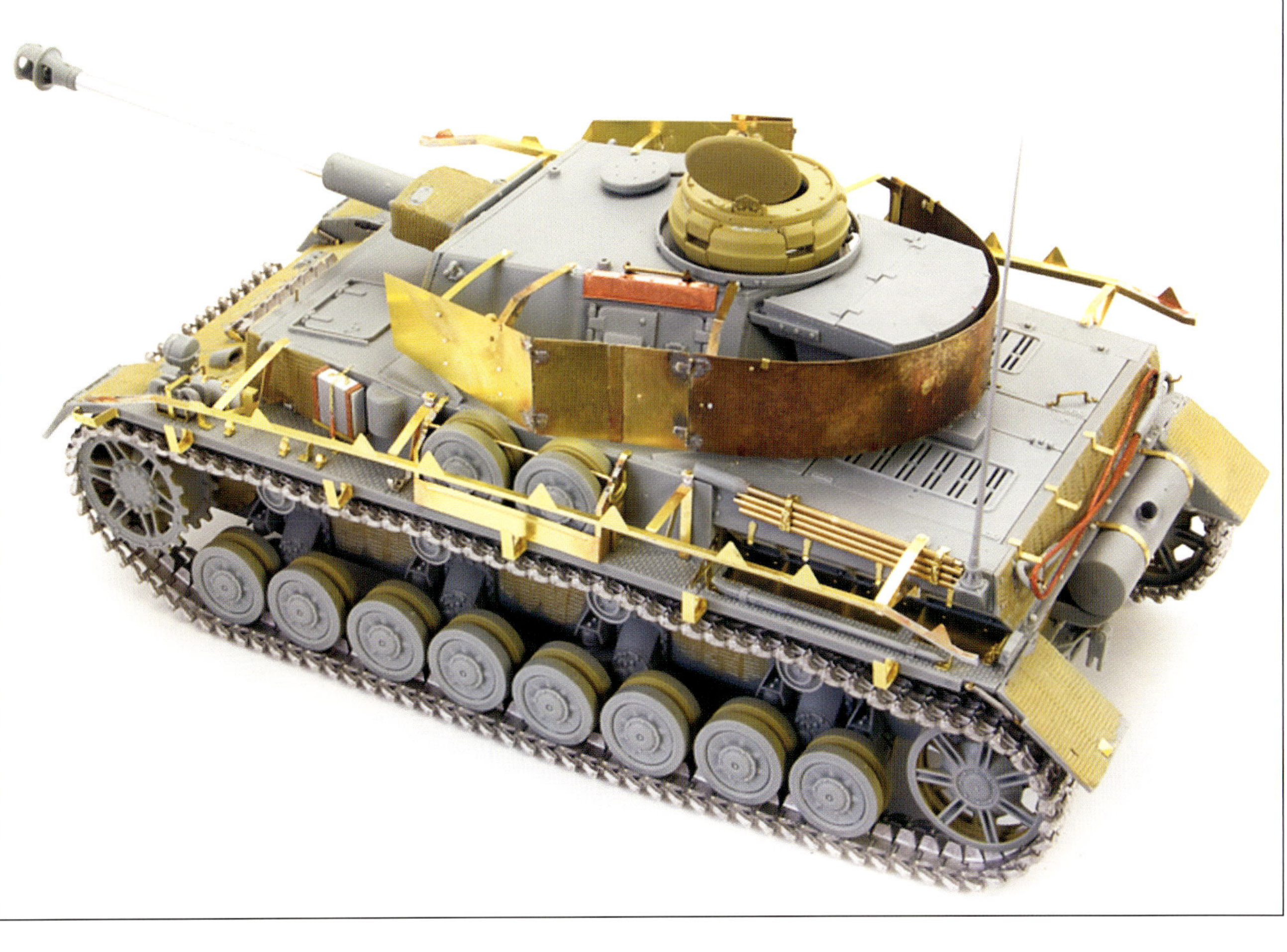

Kunihito decided to complete his replica in a plain coat of RAL 7028 Dunkelgelb, which was the standard factory-finish in July 1944. The painting process utilised Humbrol enamels, Tamiya acrylics and washes from AK Interactive and involved no less than sixteen separate steps. A graphite pencil was used to simulate worn, bare metal.

Another view of Kunihito's completed model. The weathering effects range from an overall coating of dust and dirt to more serious wear, such as the scratches along the side of the turret Schürzen.

Below: Another view of the completed model with details shown at the bottom of the page. 1 The turret Schürzen with its retaining bolts and bracket with turret lifting hook in the background. 2 The turret side access hatches, Schürzen doors in the opened position and wire mesh insert, the latter seldom seen in photographs of the full-size vehicle. 3. The armoured extractor fan cover and metal plate which was fitted before adequate supplies of the Nahhverteidigungswaffe close-defence weapon became available. 4. The badly-worn cylindrical exhaust muffler and towing cable. 5. The antenna for the FuG 5 radio and its insulator. 6. Zimmerit texture around the hull front, the Kugelblende mount and radio operator's hatch.

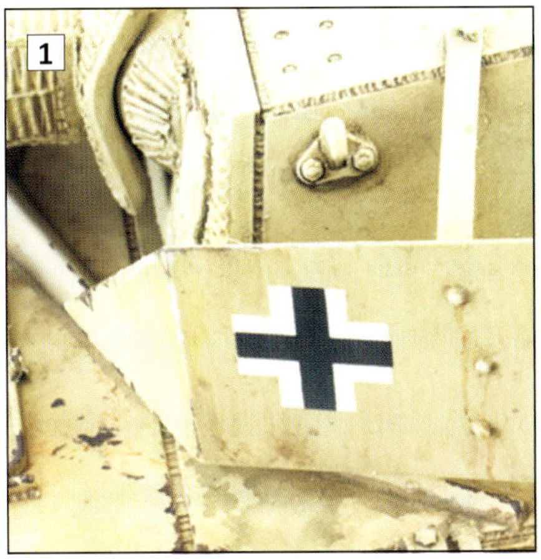

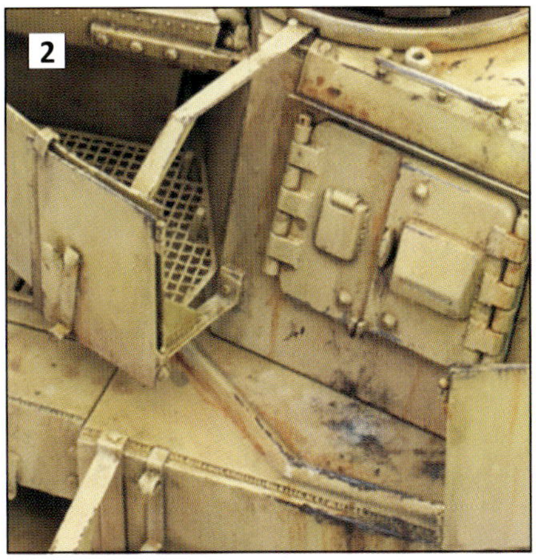

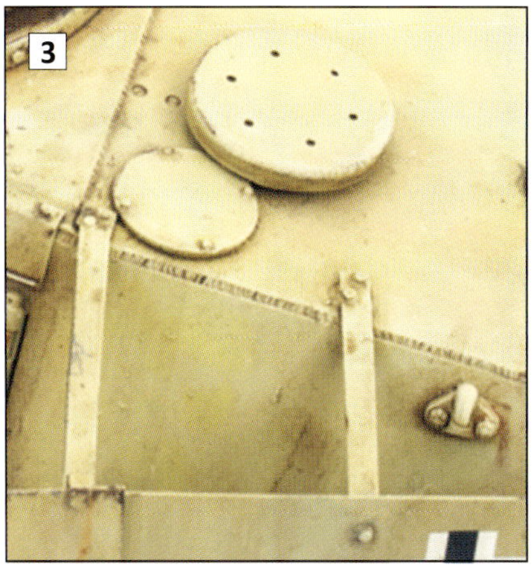

PANZER IV

AUSF H LATE PRODUCTION

PETER YOON

1/35 SCALE

Peter's replica is based on the Dragon Models Pzkpfw IV ausf H mid production kit shown at left. Below: The completed model with commander figure in place. At the bottom of the page : 1. The commander's cupola showing some of the interior detail. 2. The Filzbalg oil pre-filter. 3. Part of the Series II Schürzen rail and the vehicle's jack in its brackets on the trackguard. 4. The Kugelblende machine-gun mount. 5. A close-up view of the tank commander figure. 6. The driver's armoured visor, Bosch headlight and, just visible, the air intake of one of the brake access hatches.

Below: A rear view of the completed model showing the three spare track links, towing cable, exhaust muffler and towing coupling. Above: Detail shots of Peter's finished model including, 1. The main gun cleaning rods and antenna insulator. 2. The cylindrical exhaust muffler. 3. The rear idler and tracks with ice cleats. 4. The Bosch headlight with its cable leading into the hull. Note the Schürzen rail in the background and the method by which the plates were held in place.

Peter used Tamiya acrylic paints to create the vehicle's camouflage scheme and once that was dry a fine pin wash, made up of thinned Windsor and Newton oil colours, was applied to the surface recesses. The model was then treated with increasingly lightened shades of the main camouflage colours, dry-brushed over the entire surface.

The paint chips and scratches were built up over a considerable period, again using Tamiya acrylics and Windsor and Newton oils. The weathering, although extensive, is very realistic and suggests a tank that has been in service for some time.

The Panzerkampfwagen IV was numerically the most important tank in service with the German Army during the Second World War and it should come as no surprise that scale replicas of the various marques, from the Pzkpfw IV ausf A of 1937 to the final version of the Pzkpfw IV ausf J which left the assembly lines in April 1945, are available to the modeller. These vary in size from tiny 1/144 scale miniatures to huge 1/6 scale motorised models and includes highly detailed and accurate replicas as well as basic, snap-together kits. To list all these would not be possible in the space available and I have limited this section to the major manufacturers of 1/35 scale kits with an acknowledgement of the various 1/48 scale and 1/72 scale offerings. For the same reasons, I have tried to restrict this list to the versions most commonly encountered during the late war period, that is to say the Pzkpfw IV ausf H and the Pzkpfw IV ausf J. Similarly, I have only made brief mention of the many smaller, specialised companies which produce additional items in resin, photo-etched metal and other materials that allow a level of super-detailing which is almost limitless. I should mention that I have limited any comment or critique of these kits or products to those of which I have first-hand knowledge. An index of manufacturers can be found on page 64.

TAMIYA INC

This Japanese company has been manufacturing plastic kits since 1959, releasing their first armour model two years later. By the early 1970s Tamiya was producing a large range of armoured vehicles and accessories and the company was almost solely responsible for the rise in popularity of 1/35 scale. Tamiya's early kits were, for their day, highly detailed, relatively easy to assemble and, most importantly, affordable. The company's 1/35 scale Pzkpfw IV ausf H was first released in 1975 but has been upgraded since then and the Pzkpfw IV ausf J is a completely new kit. From 2003 Tamiya began releasing a series of armour models in 1/48 scale and the current catalogue includes a late production Pzkpfw IV ausf H and an early production Pzkpfw IV ausf J. The Trumpeter 1/16 scale Pzkpfw IV is a re-boxing of the motorised Tamiya kit that was originally released in 2008.

Above: Tamiya's 1/35 scale early production Pzkpfw IV ausf J. Below: The 1/16 scale Pzkpfw IV ausf J. First released in 2008 and upgraded in 2017, this expensive kit is fully motorised.

DRAGON MODELS LTD

At the time of writing this Hong Kong-based company's catalogue contained over thirty models of the Pzkpfw IV in 1/35 and 1/72 scale including vehicles which were in service during the last months of the war. It should be noted that a number of these are repackaged kits, some containing additional parts or markings, and a number are pre-painted display models. Nevertheless, this is an impressive total and the Dragon kits of the Pzkpfw IV ausf H and early model Pzkpfw IV ausf J are highly detailed replicas containing optional parts, photo-etched elements and a choice of markings. To complement the vehicles there is a large range of highly detailed, plastic multi-pose figures.

Above: Dragon Models' Pzkpfw IV ausf J early production model with Zimmerit. At right: The last production version in 1/35 scale. Both vehicles are also shown in the Camouflage & Markings section.

At left: Dragon Models' Pzkpfw IV ausf J last production model completed with the upgrade set from E.T. Model. Below: Details of the same model. 1. The turret roof with commander's cupola and final pattern fume extractor, cut to accommodate the Nahverteidigungswaffe close defence weapon. 2. The poison gas detection panels and wire mesh below the stowage box. 3. The Flammentöter exhaust mufflers.

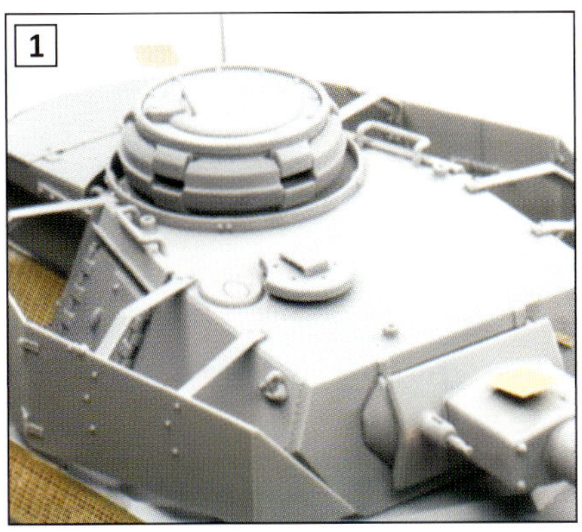

MINIART CREATIVE MODELS

Based in Ukraine, this company has been producing 1/35 vehicles, figures and aircraft models since 2001. The current catalogue contains a number of models based on the Pzkpfw IV ausf G, Pzkpfw IV ausf H and Pzkpfw IV ausf J and all include photo-etched brass parts and extensive marking options. The Krupp-Grusonwerk kit shown below contains photo-etched parts, individual track links, with a jig to help with assembly, and extensive interior detailing.

Above: Miniart's Pzkpfw IV ausf J and Panzerbeobachtungswagen IV models, both of which feature the production modifications covered by this book.

At right: Miniart's 1/35 scale Pzkpfw IV ausf G kit. Although this model depicts a mid-1943 production vehicle many of these early versions were still in service in 1945, with the Nebelwurfgerät removed from the turret side of course. This image also shows the Series I Schürzen clearly.

TRUMPETER PLASTIC MODEL KITS/HOBBY BOSS

Often thought to be separate entities, these Chinese companies are in fact owned by the same corporation, are located at the same address and are essentially the same firm although different models are marketed separately under the two labels. In 2015 Trumpeter released a 1/16 scale Pzkpfw IV ausf H and an early and later version of the Pzkpfw IV ausf J, both of which are shown below. Hobby Boss produces a series of early Pzkpfw IV tanks armed with the short-barrel 7.5cm KwK L/24 gun, all in 1/35 scale, a number of which were pressed into service in 1945.

ACADEMY PLASTIC MODEL

The company's Pzkpfw IV ausf H was originally released in 1984 and over the next twenty years was re-boxed and re-released several times, also under the Minicraft label. The latest version is a completely new kit and includes optional Zimmerit texture and although the tracks are the single-length rubber variety they are reasonably well detailed. The three marking options represent tanks of SS-Panzer-Regiment 12 and Panzer-Lehr-Regiment 130 and would not be appropriate for a vehicle of the 1945 period. The Pzkpfw IV ausf H kit released by Airfix in 2018 and the 2019 offering from Revell are the same model with different markings. The promised Pzkpfw IV ausf J seems to have been replaced by a late-production Pzkpfw IV ausf H, released in October 2022.

At left: Academy's 1/35 scale late-production Pzkpfw IV ausf H showing the Series II Schürzen fittings and the wheels and tracks before painting

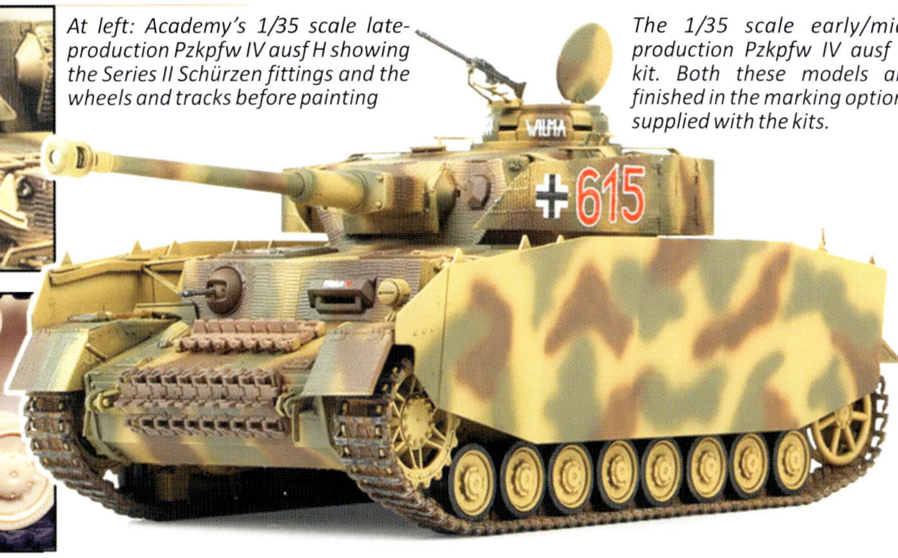

The 1/35 scale early/mid-production Pzkpfw IV ausf H kit. Both these models are finished in the marking options supplied with the kits.

RYE FIELD MODEL

A relatively new model company based in Hong Kong, Rye Field Model's current catalogue includes an early production Pzkpfw IV ausf H and a Pzkpfw ausf J which could be built as a vehicle which participated in the last battles of the war. The latter includes extensive interior detailing and neither include Zimmerit texture which is rarely seen during the last months of the conflict.A Pzkpfw IV ausf G and a Pzkpfw IV ausf G/H were to be released in 2021 but do not appear to be available at the time of writing. The company also produces sets of workable tracks.

Rye Field Model's 1/35 scale Pzkpfw IV ausf J. This is the version with limited interior detail. The company has also recently released a number of detail sets which can be bought separately from the model kits. Shown here are: 1. Photo-etched brass front fenders with hinges. 2. Detailed hull Schürzen supports and brackets, also in photo-etched brass with brass rod.

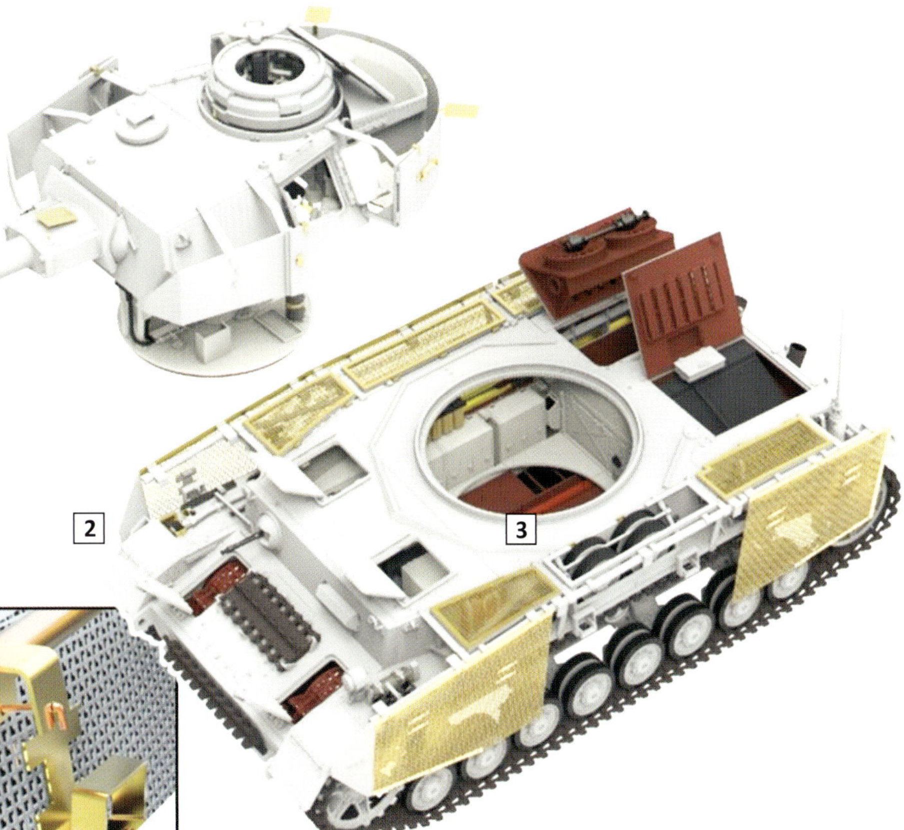

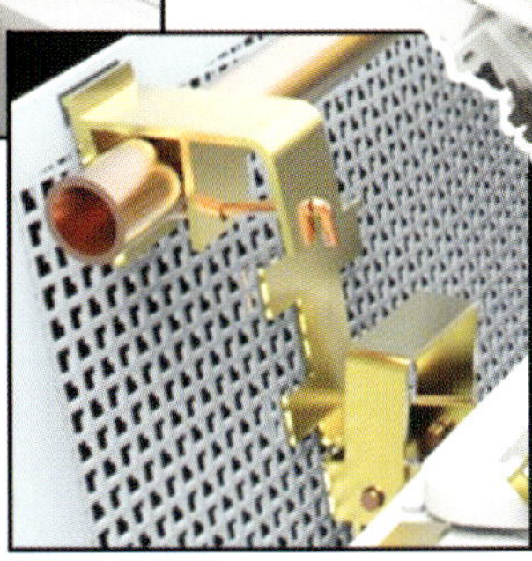

BORDER MODEL

Situated just outside Shanghai, this company began releasing kits based on the Pzkpfw IV chassis and hull in 2019, the latest offering being a reworked 1/35 scale Möbelwagen 3.7cm Flakpanzer. Border Model also manufactures workable tracks and photo-etched brass detail sets for their replicas. The Pzkpfw IV ausf H, Pzkpfw IV ausf J and Panzerbeobachtungswagen IV models, the most relevant for our period, contains crew figures, photo-etched parts, a metal gun barrel, a tool to inscribe the Zimmerit texture and fourteen marking options.

At left: Border Model's 1/35 scale Pzkpfw IV ausf J late-production model. The kit contains photo-etched brass parts and a turned aluminium gun barrel. Below: The 1/35 scale Panzerbeobachtungswagen IV model built with the parts included in the kit.

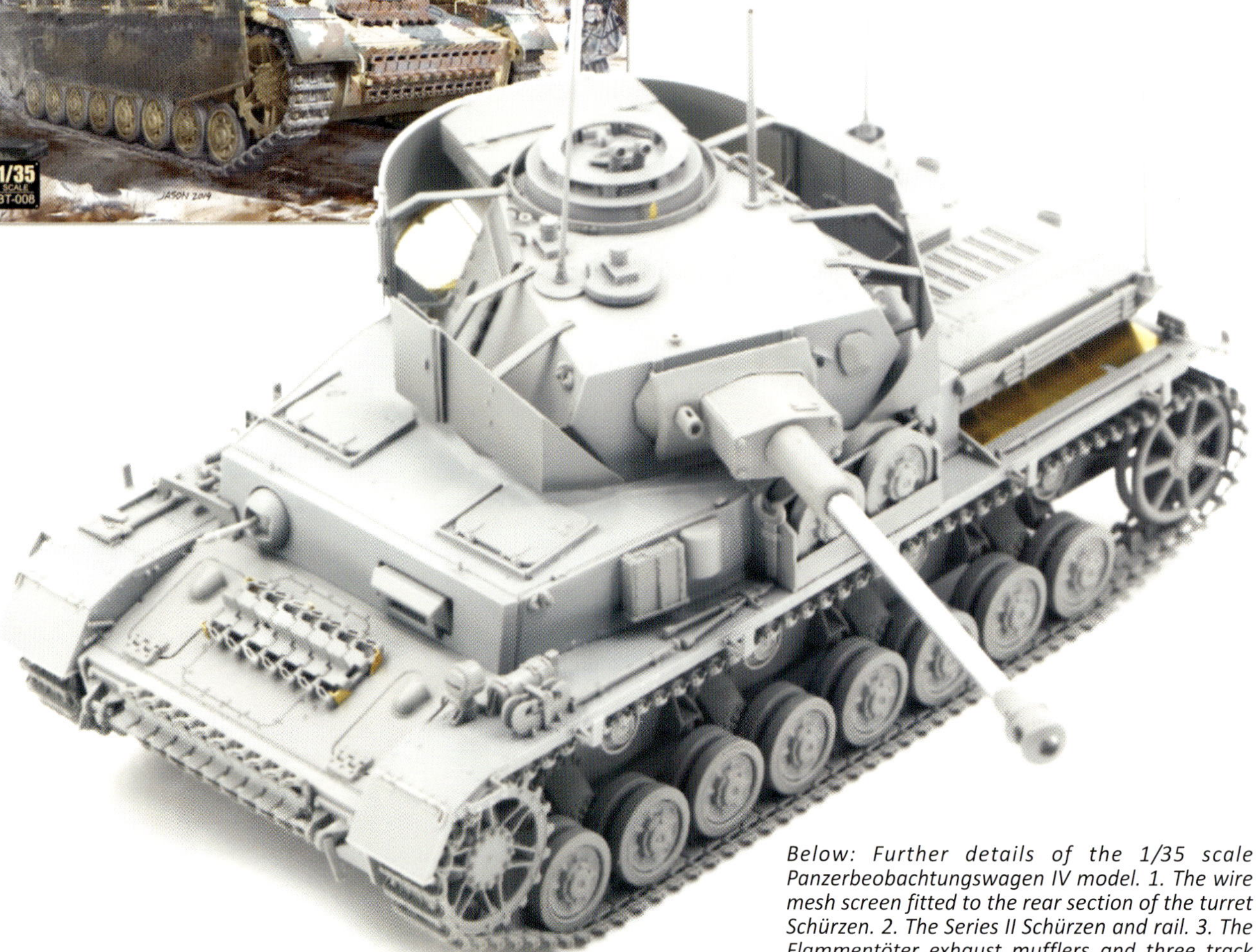

Below: Further details of the 1/35 scale Panzerbeobachtungswagen IV model. 1. The wire mesh screen fitted to the rear section of the turret Schürzen. 2. The Series II Schürzen and rail. 3. The Flammentöter exhaust mufflers and three track hangers on the hull rear plate, a feature of Nibelungenwerk-produced vehicles.

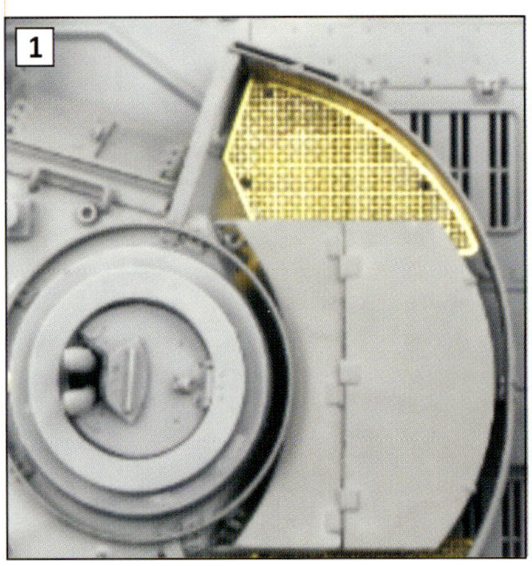

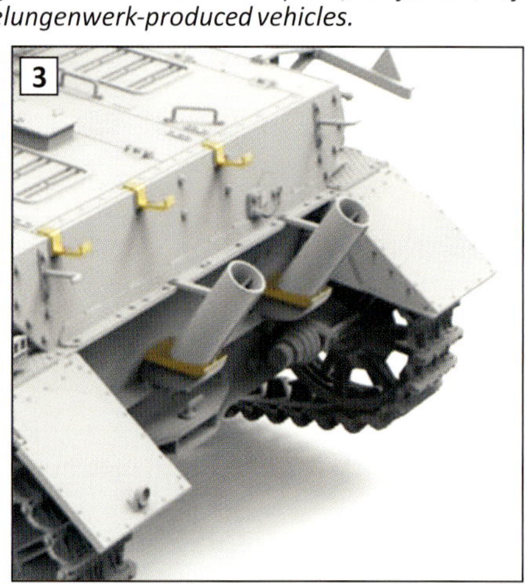

REVELL & AIRFIX

Any list of scale model tanks should include at least some mention of these two pioneering companies. The Airfix 1/76 scale kit, marketed simply as Panzer IV, was first released in 1971 and although the box art may have been altered many times over the years, the kit itself remains unchanged. It could only be described as basic but did fill a very large gap when it was released. The 1/35 scale model marketed as Panzer IV Ausf.H Mid Version is the Zvedzda Models kit which was originally released in 2017. Like Airfix, Revell has gone through many changes of ownership over the years since its founding in 1943, at one time operating as two distinct entities in the US and Germany. The company's current catalogue includes a 1/35 Pzkpfw IV ausf H which is in fact a re-boxing of Zvezda's 2017 model. The company's two 1/72 scale releases of a Pzkpfw IV ausf H and a Pzkpfw IV ausf J are reasonably accurate and easy to assemble.

At left: Airfix's 1/76 scale Panzer IV kit. This model comes with the first pattern muzzle brake fitted to the early-production Pzkpfw IV ausf G models. Below: Revell's 1/35 scale Pzkpfw IV ausf H. At far left: The marking options included with the Revell kit.

ITALERI

This Italian company has been producing plastic models since its foundation in 1962 and the company's 1/35 scale Pzkpfw IV ausf H kit is an old model that has been re-boxed under various labels including the now-defunct Testors and Tomy, as well as Zvezda and Revell. The latest incarnations were re-released in 2012 and 2012 with new parts including photo-etched details and a metal gun barrel. The 1/72 scale offering is the ESCI model first released in 1974 and then again in 1988 with individual track links. Although old, this is a reasonably good kit and it should be said that ESCI models were somewhat groundbreaking for their day. A number of Italeri's armour models have also been released under the World of Tanks label and also as part of the Girls und Panzer franchise.

Italeri's 1/35 scale Pzkpfw IV ausf H built with the photo-etched brass detail set from Eduard Model Accessories. Many of these vehicles, as depicted here, were in service until the end of the war with the exception of the ungainly Filzbalg oil filter on the hull side which was dropped from production in early 1944.

AFTERMARKET DETAILS

The manufacturers listed in the previous pages are the major players in this industry but there are many smaller companies which specialise in upgrade sets in photo-etched brass, aluminium, resin and wood to enhance the injection-moulded kits. Readers of this series will already be familiar with the more well known companies including Eduard Model Accessories, Aber and RB Model from the Czech Republic and Poland, Voyager Model and Griffin Model from China, DEF.Model which is based in Korea and Passion Models from Japan.

Any attempt to provide a definitive list would be impractical and, as representative, I have highlighted here the products of E.T. Model, a relatively new manufacturer from China, which produces upgrade sets in 1/72 and 1/35 scale in brass and resin. Most sets are specifically designed, according to the company, to fit Dragon or Tamiya kits but some are referred to as universal. The staff at E.T. Model are all hobbyists and the outstanding work of one, Penny Shi, appeared in the title on the Sturmgeschütz III and Sturmgeschütz IV.

At right: E.T. Model's photo-etched and resin upgrade set applied here to Dragon Models' 1/72 scale Pzkpfw IV ausf G. Below: The highly-detailed photo-etched brass set in 1/35 scale for the last-production Pzkpfw IV ausf J with a complete set of wire mesh screens and support rails.

Below: 1. Further detail of the 1/35 scale last-production Pzkpfw IV ausf J showing the poison gas detection panels and Orterkompass stand fitted to the turret fume extractor. 2. The Series II Schürzen rails and brackets in 1/35 scale. 3. The same model with the Schürzen attached.

In this section of *TankCraft 43: Panzer IV Medium Tank, German Army and Waffen-SS, Normandy Campaign Summer 1944* we concentrated largely on the technical aspects of the Pzkpfw IV ausf H which was the type most commonly encountered during that time period. By the end of 1944 the Pzkpfw IV ausf J had been in production for approximately ten months and it is the modifications that were introduced with this version that we will examine in detail here. But to provide some context and present a more coherent picture, mention should be made of earlier developments as many of these were incorporated into the new design. In addition, significant numbers of earlier models remained in service until the end of the war and a number of these are shown the Camouflage & Markings section which begins on page 17.

The Pzkpfw IV ausf H was assembled by the firms of Friedrich Krupp AG, Vogtländische Maschinenfabrik (Vomag), which built the first examples, and at the massive Nibelungenwerk (NiWerk) at Sankt Valentin in Austria. The first vehicles left the assembly lines in May 1943 and the early production models of the new tank were externally similar to its predecessor, the Pzkpfw IV ausf G. The distinctive welded or bolted-on armour continued to be fitted to the front of the hull and fighting compartment but the visors on the hull sides were dropped from production and the turret roof was strengthened. The radio antenna was moved from its folding position on the hull right side to the left rear side where the insulator was held in a metal bracket. As protective hull and turret Schürzen, or skirts, were incorporated into production from as early as April 1943, obscuring many details, the position of the antenna is a useful indicator. The wooden trough which protected the folded antenna of the Pzkpfw IV ausf G was removed from the hull side and its place taken by a Filzbalg oil-filter but again this is often difficult to discern with the hull Schürzen in place. The new tank was fitted with a reinforced final drive and this necessitated a redesigned, welded drive sprocket which was fitted to all vehicles except the first thirty tanks assembled by Vomag. One distinguishing feature of NiWerk-produced tanks was the three spare track links held on the hull rear plate on the left side.

All vehicles were allocated a Fahrgestelle, or chassis, number abbreviated here as Fgst. Nr. The major modifications incorporated into the production of the Pzkpfw IV ausf H and Pzkpfw IV ausf J are listed below.

April 1943. Before assembly of the Pzkpfw IV ausf H could begin the Krupp plant was ordered to abandon its manufacture and to prepare for production of the Panther I or Panther II (1) which were then in the development stage. In Pzkpfw IV ausf G production the longer 7.5cm L/48 gun replaced the L/43 version and a redesigned multi-baffle muzzle brake was introduced. This weapon was a standard feature of the Pzkpfw IV ausf H. As a defence against Soviet anti-tank rifles, armoured skirts, referred to as Schürzen, were fitted to the turret and hull sides. The development of the Schürzen, particulary as it affected the Pzkpfw IV ausf J, is dealt with in detail in the section which begins on page 56.

May 1943. The first Pzkpfw IV ausf H tanks left the assembly plants in this month. For the entire project Fgst. Nr. were allocated in blocks to the three major manufacturers in the range 84401 to 90600. During May the managers at Krupp were notified that the Panther II project would be terminated and that some Panther I should be completed as Bergepanther recovery tanks.

..........*text continued on page 51*

Notes

1. The Panther I is more commonly referred to today as the Panther ausf D and the rather complicated history of the naming of these tanks is explained in detail in *TankCraft 34: Panther Medium Tank, German Army and Waffen-SS, Eastern Front Summer 1943.*

This late-production Pzkpfw IV ausf G, photographed on the Eastern Front, is fitted with a number of the modifications mentioned in the text. A. The distinctive Filzbalg pre-filters which were probably only fitted to this version during the last month of its production run. B. The Zusatzpanzer armour welded to the hull front. C. The support rails for the Series I Schürzen plates. Note that the front towing couplings seem to have been cut in half (D).

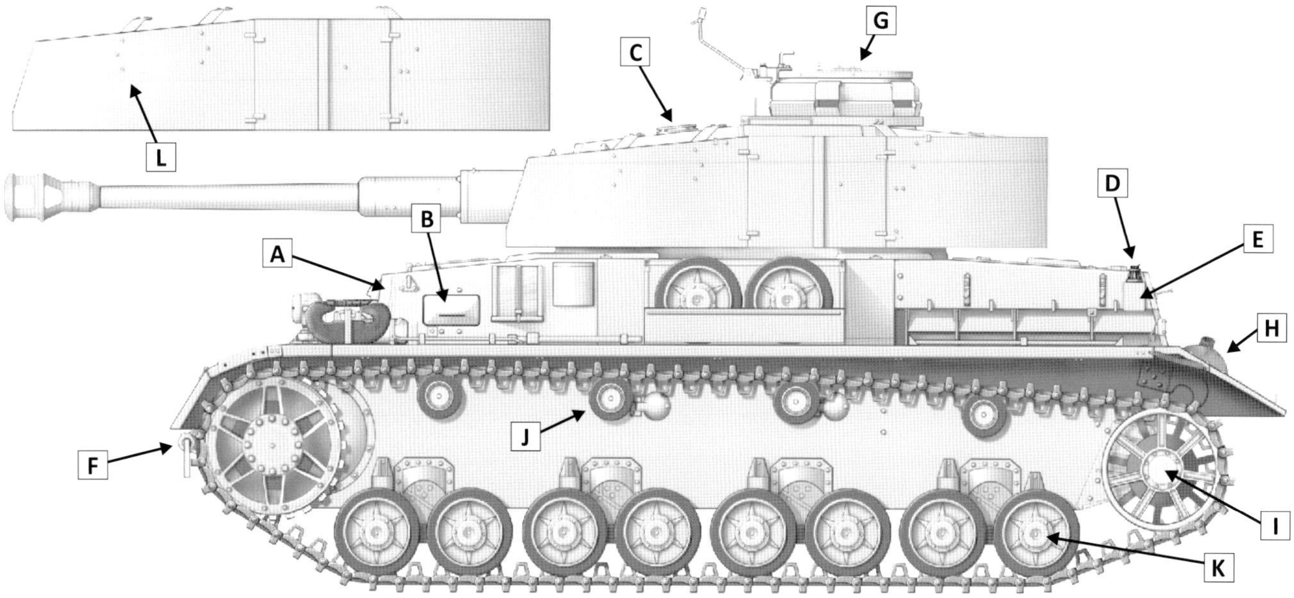

Above: Pzkpfw IV ausf H, typical of vehicles built by Krupp during the autumn of 1943 after the Series II Schürzen had been introduced into production. I have deliberately omitted the hull Schürzen plates on this and the following drawings for the sake of clarity and these are examined in detail on pages 56 and 57. By this time the armour contractors were supplying hulls with integral 80mm front plates (A) and without the side vision ports. I have left the latter in place (B) as a reference and in fact they could still be seen for some time. Note the fume extractor cover (C) on the turret roof, the antenna insulator (D) and its bracket (E) on the left side at the rear which was moved to that position with the first Pzkpfw IV ausf H. Just visible is the bolted-on front towing coupling (F), the hinge for the hatch of the commander's cupola (G), which opened towards the left, and the cylindrical exhaust muffler (H). The cast rear idler (I) was fitted to the first tanks in October 1943 and at about the same time the rubber return rollers (J) were being phased out in favour of the metal versions and cast hub caps were fitted to the roadwheels (K). The arrangement of the bolts (L) holding the turret Schürzen in place on tanks assembled by Vomag was slightly different. Also omitted from this drawing is the coating of Zimmerit anti-magnetic mine paste which was a standard feature from September 1943.

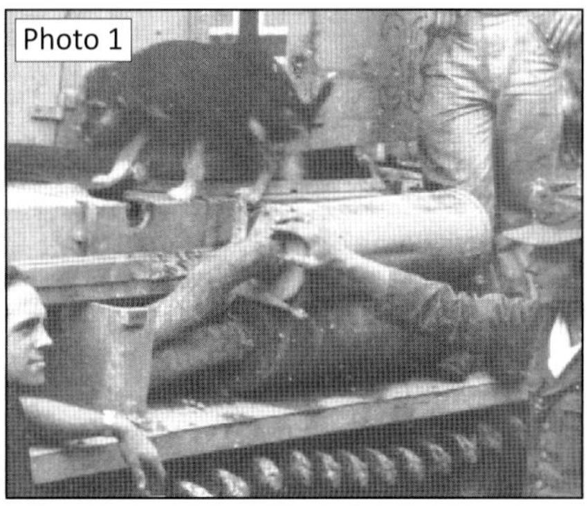
Photo 1

Photo 2

Photo 3

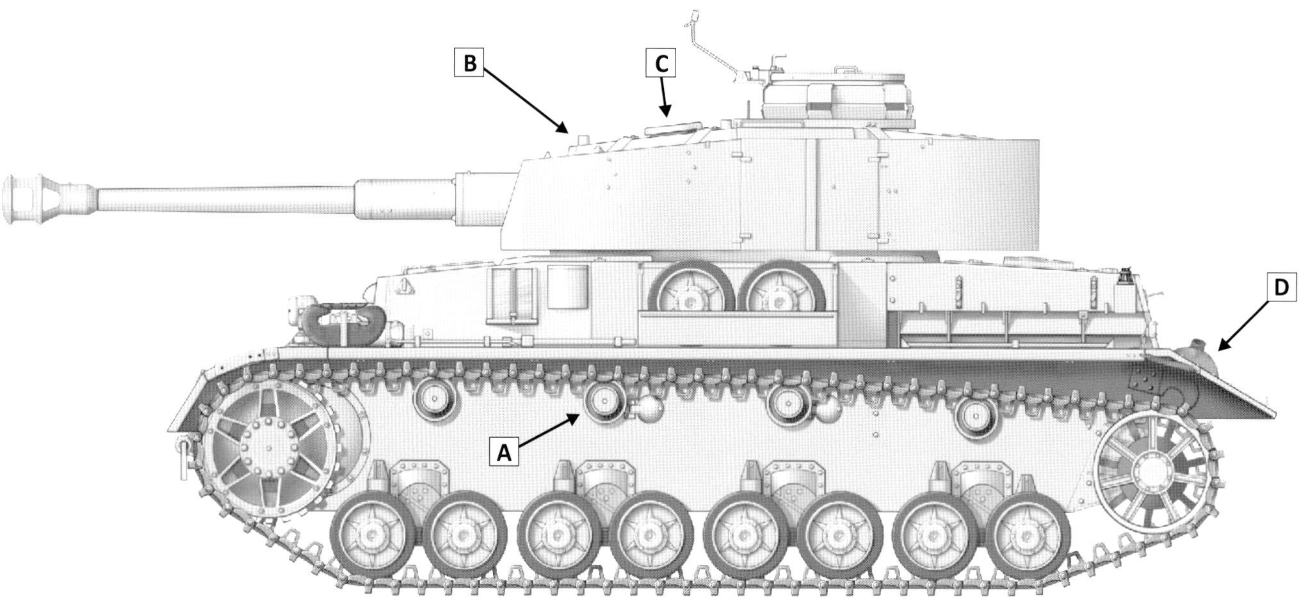

Above: Pzkpfw IV ausf J built at the Nibelungen plant during the summer of 1944. The most obvious differences between this model and the Pzkpfw IV ausf H were the elimination of the Filzbalg pre-filter fitted to the hull side and the 2-stroke turret traverse engine positioned next to the exhaust muffler (photo 1 and photo 2, above). The side plates of the turret Schürzen were a distinctive feature of tanks assembled by this manufacturer as the sides were mounted 50mm further forward, which necessitated changes to the doors and the support brackets. In addition, three claps were welded to the rear of the engine compartment for spare tracks (photo 3, above). Note the rubber-saving steel return rollers (A), or Gummilose Stützrolle, Pilze sockets (B) and armoured cover for the fume extractor (C). The cylindrical exhaust (D) was replaced by Flammentöter mufflers in August 1944.

..........text continued from page 49

June 1943. The last Pzkpfw IV ausf G tanks were assembled in this month, many incorporating features of the Pzkpfw IV ausf H. Nebelwurfgerät, smoke candle dischargers, were no longer fitted to the turret sides (1) as it was believed that small arms fire could cause them to prematurely ignite. A bracket for an anti-aircraft MG 34 machine gun was fitted to the commander's cupola.

Beginning in June the hull side vision ports were to be gradually dropped from production and completely phased out by September. Zusatzpanzer, or additional armour, in the shape of 30mm thick plates, had been welded or bolted to the hull glacis and superstructure front in the Pzkpfw IV ausf G production run from May 1942, gradually becoming standard in February 1943. The welded plates were dropped from the manufacture of the Pzkpfw IV ausf H at the end of May in favour of the bolted-on type and both were superseded by a homogenous 80mm-thick armour plate in June 1943. This is, however, rather simplistic and vehicles could be seen with a mix of all three.

August 1943. The side visors and rear pistol ports on the turret were dropped from production. Krupp was ordered to halt all Panther I assembly and begin production of the Pzkpfw IV, with a monthly target set at 150. Later the same month the target was dropped to 100.

September 1943. Late in the month the factories began applying Zimmerit anti-magnetic mine paste to the vertical armour surfaces. For a short period Zimmerit was also applied to the hull and turret Schürzen and the horizontal surfaces around the driver's and radio operator's hatches. In the same month rectangular splash guards were introduced for those hatches.

October 1943. The rubber-tyred return rollers were replaced by cast steel versions. A cast metal rear idler replaced the welded, tubular version that had been standard since 1941 with the manufacture of the Pzkpfw IV ausf F. Both these changes were introduced gradually and Krupp in particular was slow to fit the new return rollers and rear idler.

November 1943. Three spare track links were fitted to the track guards. Krupp-built vehicles had two track links on the right front track guard and one behind the spare wheel bin on the left side.

December 1943. To strengthen the hull, the 80mm-thick front plate was interlocked with the hull side extensions during production.

An order for Krupp to assemble 100 Sturmgeschütz IV assault guns per month was rescinded and the company directed to build 150 Pzkpfw IV per month. A contract with the same firm to produce the

Notes

1. Photographic evidence shows that these were indeed included in early Pzkpfw IV ausf H production.

A Pzkpfw IV ausf J of Panzer-Regiment 15 photographed in western Germany, possibly in the Ruhr area, in early 1945. Just visible on the hull front is the famous Gespenster unit insignia of 11.Panzer-Division, the regiment's parent formation. An enlarged version is shown at right. This vehicle is fitted with the pivoting cupola hatch incorporated into production in July 1944 but lacks the Pilze sockets introduced at about the same time. The absence of Zimmerit would suggest that this tank was assembled at some time after September and it may be one of the fifteen Pzkpfw IV tanks delivered on the last day of November 1944.

Notes

1. HWA, the Heereswaffenamt, was the army's ordnance office responsible for the design, testing, development, and acceptance of weapons and equipment.

Panzerbefehlswagen IV was cancelled in January 1944, causing the company's management to later claim that this kind of indecision accounted for 300,000 lost man-hours which equated to 150 tanks.

January 1944. A Nahverteidigungswaffe, or close-defence weapon, was supposed to have been mounted in the turret roof but a shortage of weapons meant that many tanks left the factories without it and it was not a standard feature until July 1944. On vehicles where the necessary aperture had been cut in the turret roof, the hole was covered by a metal plate secured by four bolts. Units in the field were ordered to apply Zimmerit to their tanks.

February 1944. An improved oil filter, installed in the engine compartment, replaced the Filzbalg filters in production. Units in the field were ordered to remove the Filzbalg filters from their tanks but many were still in place during the summer battles. Zimmerit was now applied to the hull sides as well as the other vertical surfaces. A programme of converting Pzkpfw IV ausf G tanks returned to Germany for repair to Panzerbefehlswagen command tanks began in this month.

In February 1944 the first Pzkpfw IV ausf J vehicles were completed and accepted by the HWA (1) and by the following month had completely superseded the Pzkpfw IV ausf H in production. Assembly was undertaken by Nibelungenwerk and Vomag and the two firms completed about 3,160 vehicles in the Fgst.Nr. range 86394 to 110272 by April 1945.

The major external difference between the new version and earlier models was the

..........*text continued on page 52*

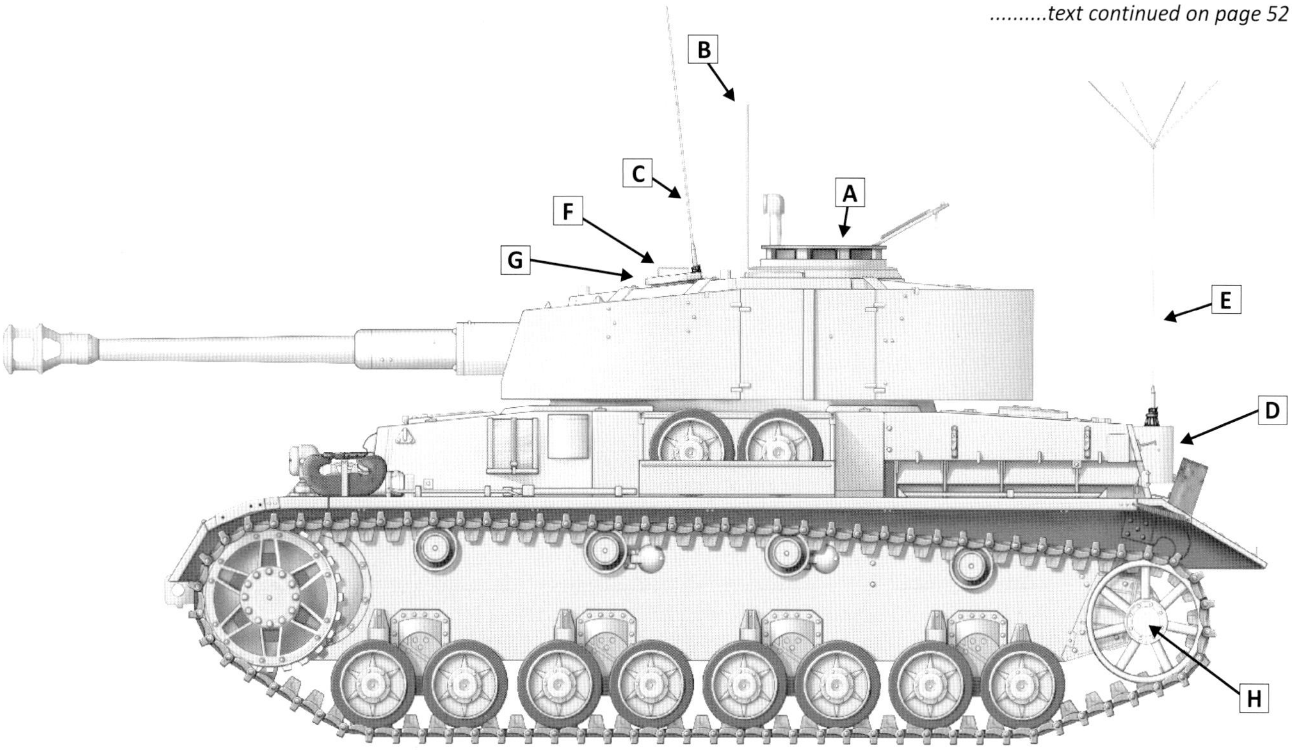

Panzerbeobachtungswagen IV ausf J assembled by Nibelungenwerk during the autumn of 1944. Standard tanks were taken from the Nibelungenwerk production line for conversion to these specialist artillery observation vehicles beginning with ten in April 1944. There was a considerable delay before the first conversions were completed and accepted by the HWA and it was not until July that any tanks were allocated to operational units. The monthly figures steadily increased, except in February 1945 when none were built, and by the end of the war a total of 140 Panzerbeobachtungswagen IV had been completed and accepted. The most distinctive feature of these vehicles was the Sturmgeschütz cupola (A) fitted with seven periscopes. A T.S.R. 1 periscope (B) was fitted in a pivoting mount in the turret roof and each vehicle was supposed to be equipped with an Orterkompass although it is unknown if any were issued. The 7.5cm main gun was fully functional but ammunition stowage was decreased and the mount for the coaxial machine gun was removed in order to accommodate the extra radio equipment. The intention was that these vehicles should be allocated to the commanders of self-propelled artillery batteries and most authoritative accounts state that each tank was fitted with a Funkgerät (FuG) 4 medium-wave receiver, an FuG 8 medium-wave transceiver and a Funksprechgerät f, a radio system normally reserved for Panzergrenadier units. But this arrangement has recently been questioned and contemporary documents and operating instructions suggest that the radios were the FuG 5, that was standard for the Pzkpfw IV, with the antenna fitted in the turret roof (C) and the FuG 8 in an armoured cover (D) on the hull rear with its distinctive Sternantenna D (E). Power for the radio sets was provided by a GG 400 electrical generator installed in the left rear of the fighting compartment, as it was for the command version, and the exhaust tube which protruded for the hull rear is another identifying feature of these tanks. Note the bracket (F) for the Orterkompass mounted on the armoured fume extractor cover (G) and the earlier-style welded rear idler (H) which was never fully replaced by the cast version.

Facing page: The Panzerbefehlswagen IV ausf J Fgst Nr.90200 of SS-Panzer-Regiment 1 which was attached to Kampfgruppe Peiper and abandoned during the fighting in the Ardennes. Points to note include: A. The anti-aircraft machine-gun bracket, here in the raised position. B. The T.S.R. 1 periscope which was fitted to these purpose-built command vehicles and also to the Panzerbeobachtungswagen IV. C. An interior view showing the T.S.R. 1 periscope. The Nahverteidigungswaffe close defence weapon was not fitted to these vehicles and it was replaced by an additional antenna mount (D) while the bracket for the Orterkompass (E) was located behind the fume extractor cover (F). The Sternantenna D, when not in use, was stored in the box (G) on the hull side. The same vehicle photographed from the rear. H. The armoured cover for the Sternantenna D aerial insulator. The frame visible here (I) was referred to as a Stützkorb, literally support basket, and allowed a 0.7m extension to be added to the aerial. Note that there is no antenna insulator fitted to the bracket (J) on the hull side. K. The exhaust for the GG 400 electrical generator. L. The vehicle's Fahrgestellnummer (Fgst Nr.).

A
B
D
E
F
G
C
I
H
K
J
L

Notes

1. Interestingly, trials conducted on captured tanks by the British showed that the manual turret traverse was as effective, and in some situations better, than the electrical drive.
2. Units in the field were ordered to fit the Pilze sockets to the turrets of their tanks in the same month.

deletion of the electrical turret drive (1) meaning that the exhaust muffler for the auxiliary generator, which was located on the left side of the engine compartment, was no longer fitted to the hull rear plate. In the same month, the three spare track links on the hull rear plate, which had been an identifying feature of NiWerk-manufactured tanks, were no longer fitted.

March 1944. The cut-out in the first plate of the hull Schürzen was dropped.

April 1944. Beginning in this month a number of Pzkpfw IV ausf J tanks were converted to artillery observation tanks (see page 52). It was sometime before any of these were accepted by the HWA and the first were allocated in July 1944 to units fighting on the Eastern Front.

May 1944. An earlier decision to delete the vision ports and machine-pistol ports of the turret side access hatches was implemented but the change was slow and tanks were assembled until the end of the war with these features. The raised housing over the radiator filler caps on the rear deck were simplified by having their sides squared-off. In this month production at Vomag was diverted to the Jagdpanzer IV after less than 200 tanks had been completed.

June 1944. S-shaped hooks for the towing cables were to replace the C-shaped versions but there is strong photographic evidence that this modification was never implemented. Three sockets, or Pilze, were welded to the turret roof to hold a 2-ton jib boom that was capable of lifting large parts, such as the tank's engine (2).

July 1944. The thickness of the superstructure roof was increased to 16mm by the addition of armour plates but a plan to replace the hatches of the driver and radio operator was postponed until old stocks were exhausted. A large armoured cover was fitted to the fume extractor on the turret roof. An earlier proposal to mount a 200-litre fuel tank in place of the auxiliary generator was begun in this month when the first was fitted to Fgst Nr.91501. But within weeks serious leaks were discovered in the tanks and installation was suspended until September.

In this month Hitler ordered that manufacture of the Pzkpfw IV at the Ni-Werk plant was to end and the firm switch to production of the Panzer IV/70(A), and ultimately designs that were still on the drawing board. Most tanks are now fitted with the Nahvertidigungswaffe close defence weapon.

August 1944. At some time before the end of the month, Ni-Werk introduced flame suppressing exhaust covers referred to as Flammentöter. At least two sizes were in use by the end of the war.

Simplified hinges were introduced for the fuel filler caps and the number of bolts on each bogie carrier was reduced. The tow cable and bracket were moved to left side. Six spare track links and a spare antenna rod were fitted on the right side.

September 1944. The HWA advised manufacturers that the application of Zimmerit was to cease. Beginning with Fgst Nr.92301, wire mesh Schürzen replaced the earlier versions which

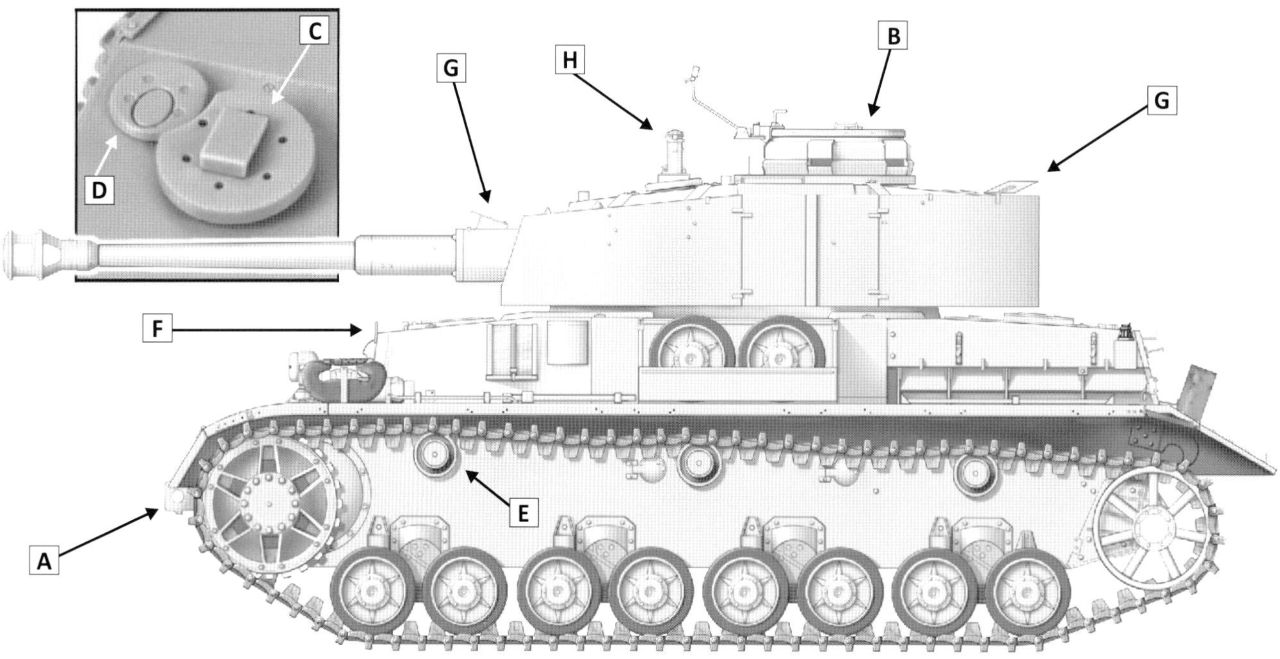

Above: Pzkpfw IV ausf J built by the Nibelungenwerk during the spring of 1945 shown with the hull support rails and brackets for the Drahtgeflechschürzen removed. From October 1944 the hull sides were extended at the front (A) and this modification replaced the bolted-on towing couplings. At the same time a pivoting cupola hatch (B) was introduced and in the following month the fume extractor cover was trimmed (C) to accommodate the Nahverteidigungswaffe close-defence weapon (D). In December the number of return rollers (E) was reduced to three and the bolted-on lifting hooks on the hull sides were replaced by simple welded loops (F). Note the Lost-Erkennungstafeln poison gas detection cards (G) incorporated into production in January 1945. I have shown the Orterkompass (H) in place although it is doubtful that any were actually issued.

consisted of metal plates. The bolted-on hinges of the driver's and radio operator's hatches were replaced by simplified versions that were welded in place.

October 1944. The hull sides were extended and eyes were cut in the extensions. This modification replaced the towing couplings that had been bolted to the hull front. Beginning with Fgst Nr.92447, a pivoting hatch was fitted to the commander's cupola. This feature was apparently introduced slowly and tanks known to have been assembled in November can be seen with the earlier version. To accommodate the Nahvertidigungswaffe a section of the extractor fan cover was cut away (1).

A mount for the Orterkompass (OK) 38 was fitted to the armoured cover. This compass was similar to those already in use with the Luftwaffe and incorporated correctors to compensate for the magnetic field created by the tank. A manual for its use with armoured vehicles was disseminated in early 1945, and its incorporation into the Tiger I programme was anticipated as early as June 1944, but it is unclear if any were actually issued.

November 1944. To simplify manufacture and strengthen the turret sides, the front vision port was eliminated from production. Turrets built with the necessary aperture were still fitted with the vision port which was now welded closed. New turrets were only gradually introduced and this change was not fully implemented by the end of the war.

December 1944. The number of return rollers was reduced to three. This change had been in the planning stage since the previous June at least. By the end of the month a new towing coupling, featuring a horizontal bolt, was fitted to the hull rear. Units in the field were supplied with the new towing coupling and ordered to replace the old version. In production, welded u-shaped brackets replaced the lifting hooks on the hull sides.

January 1945. Holders for poison gas detector cards, Lost-Erkennungstafeln, were fitted to the top of the gun mantlet and at two positions along the top edge on the rear section of the turret Schürzen. Units in the field were issued with instructions to modify the earlier hinged cupola hatch to open to about 170 degrees making it a smaller target.

February 1945. To simplify production, the air intakes on the hull brake inspection hatches were eliminated. This modification was not fully implemented before the end of the war.

March 1945. A rubber-saving radio antenna insulator, or Antennenfuss, was introduced.

April 1945. The return roller mounts which had been bolted to the hull sides were now simply welded in place.

Notes

1. It is interesting that this modification was not made until October as the armoured cover was introduced in July by which time adequate supplies of the Nahvertidigungswaffe were available.

Although this vehicle is very heavily damaged I have included this image as it shows several production features which are not often encountered in contemporary photographs. These include: A. The swivelling hatch of the commander's cupola in the open position. B. The armoured cover for the fume extractor fitted with the Orterkompass bracket (C). Note that the cover has been cut to accommodate the Nahverteidigungswaffe close-defence weapon (D), part of which is visible inside the turret (E), a modification that was introduced in October 1944. F. The brake inspection access hatches without the air-intakes dating this vehicle's production to February 1945 or later. G. The front hull extension which replaced the bolted-on towing coupling. H. The support rail (H) and adjustable brackets (I) for the Drahtgeflecht Schürzen.

A feature of German tanks and assault guns of the later-war period was a system of spaced armour referred to as Schürzen, literally skirts, which provided extra protection to the hull and in some cases the turret. Although these were introduced prior to the period covered by this study, some explanation of the development process and differences in design would be helpful. Early experiences on the Eastern Front had shown that the hull sides of the most important armoured vehicles in the army's inventory, the Pzkpfw III and Pzkpfw IV tanks and the Sturmgeschütz III assault guns, could be penetrated by the Red Army's 14.5mm PTRD and PTRS anti-tank rifles. Even the Panther, when it entered service, was susceptible to these light, man-portable weapons. Experiments at the front involving additional armour plates proved unsatisfactory and at a meeting held in February 1943 Hitler was presented with the idea of fitting tanks with spaced armour on which the enemy projectile would expend its energy. Trials were held in the following month and these proved so successful that it was ordered that Schürzen be introduced into the production process of the Pzkpfw III, Pzkpfw IV and Panther tank and the Sturmgeschütz III assault gun. Combat trials were undertaken by the companies of Panzer-Regiment 21 in May 1943 and the subsequent report suggested, contrary to earlier findings, that any advantages offered by spaced armour plates were outweighed by the system's shortcomings. Nevertheless, turret and hull Schürzen were incorporated into the assembly of the Pzkpfw IV in the same month and would remain an element of production until the war's end.

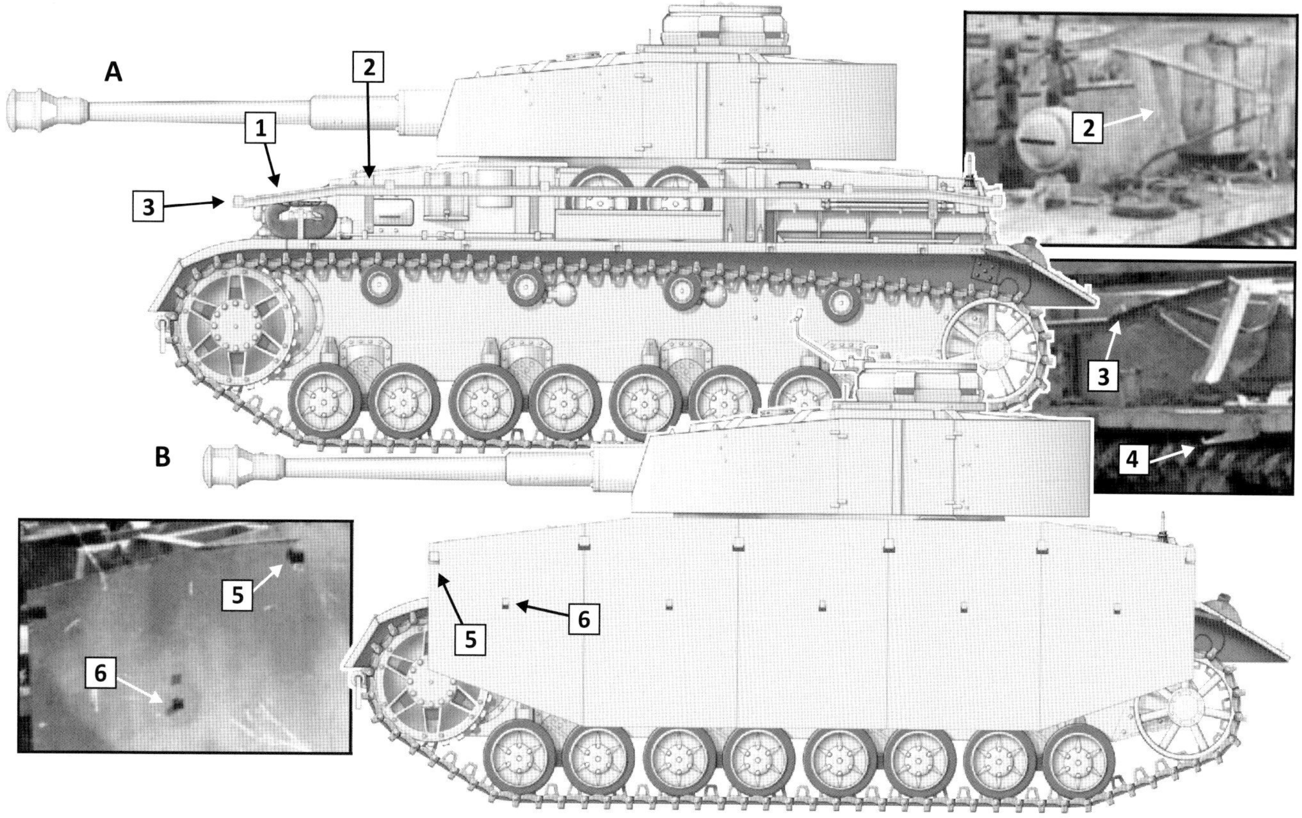

Above: A. Turret and hull Schürzen of the initial design, often referred to as Series I, were introduced into the production process in May 1943 just one month after the first Pzkpfw IV ausf H vehicles, the model shown here, left the assembly plants. The supporting rail (1) for the hull plates was held in place by four brackets (2) which were bolted to the hull sides and roof. The armour plates were suspended from six hook-like hangers on the rail (3) and five smaller hangers (4) bolted to the trackguard. B. The same Pzkpfw IV ausf H vehicle with the hull Schürzen plates in place. Note how the hangers of the supporting rail protrude between each plate (5) while the smaller hangers fitted to the trackguard (6) were intended to provide further stability.

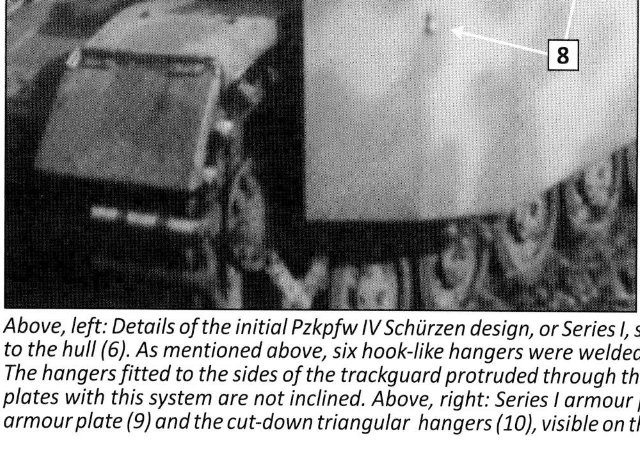

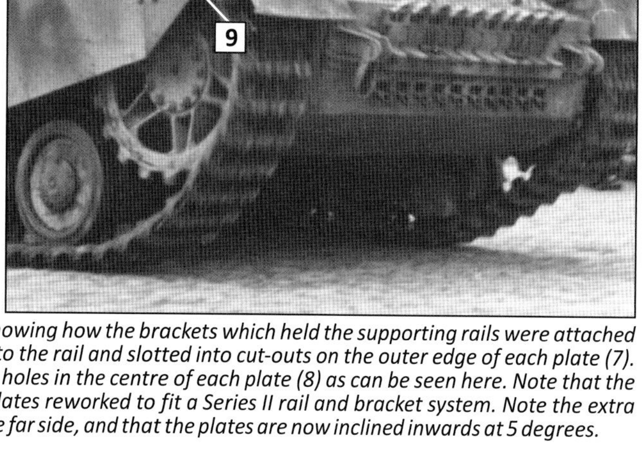

Above, left: Details of the initial Pzkpfw IV Schürzen design, or Series I, showing how the brackets which held the supporting rails were attached to the hull (6). As mentioned above, six hook-like hangers were welded to the rail and slotted into cut-outs on the outer edge of each plate (7). The hangers fitted to the sides of the trackguard protruded through the holes in the centre of each plate (8) as can be seen here. Note that the plates with this system are not inclined. Above, right: Series I armour plates reworked to fit a Series II rail and bracket system. Note the extra armour plate (9) and the cut-down triangular hangers (10), visible on the far side, and that the plates are now inclined inwards at 5 degrees.

It was very soon discovered that the Series I Schürzen arrangement presented a number of drawbacks. In addition to the rails being too weak to absorb even mild pressure, the armour plates tended to push mud into the drive train which placed excessive strain on the engine and in dry conditions dust and sand accumulated between the plates and the tank's hull. The latter concern was particularly problematic for vehicles fitted with the Filzbalg oil filter. Despite these complications, the basic idea of spaced armour was retained and from September 1943 a new design was fitted by all manufacturers which consisted of a strengthened rail (1) and brackets (2). Triangular hangers (3) replaced the smaller rectangular variants of the Series I and the hangers fitted to the trackguards (4) were also redesigned. From November 1943 the triangular hangers at the front were trimmed to conform to the top edge of the armour plates (5). The plates were essentially the same as the Series I versions but were inclined at 5 degrees to the perpendicular. An additional plate (6) was added at the front and this was angled towards the hull as can be seen in our photograph (7). U-shaped brackets (8) were welded onto the inner face of the Schürzen plates to hold the triangular hangers. Later versions of these brackets allowed the plates to be held further from the trackguard when the wider Ostketten tracks were fitted.

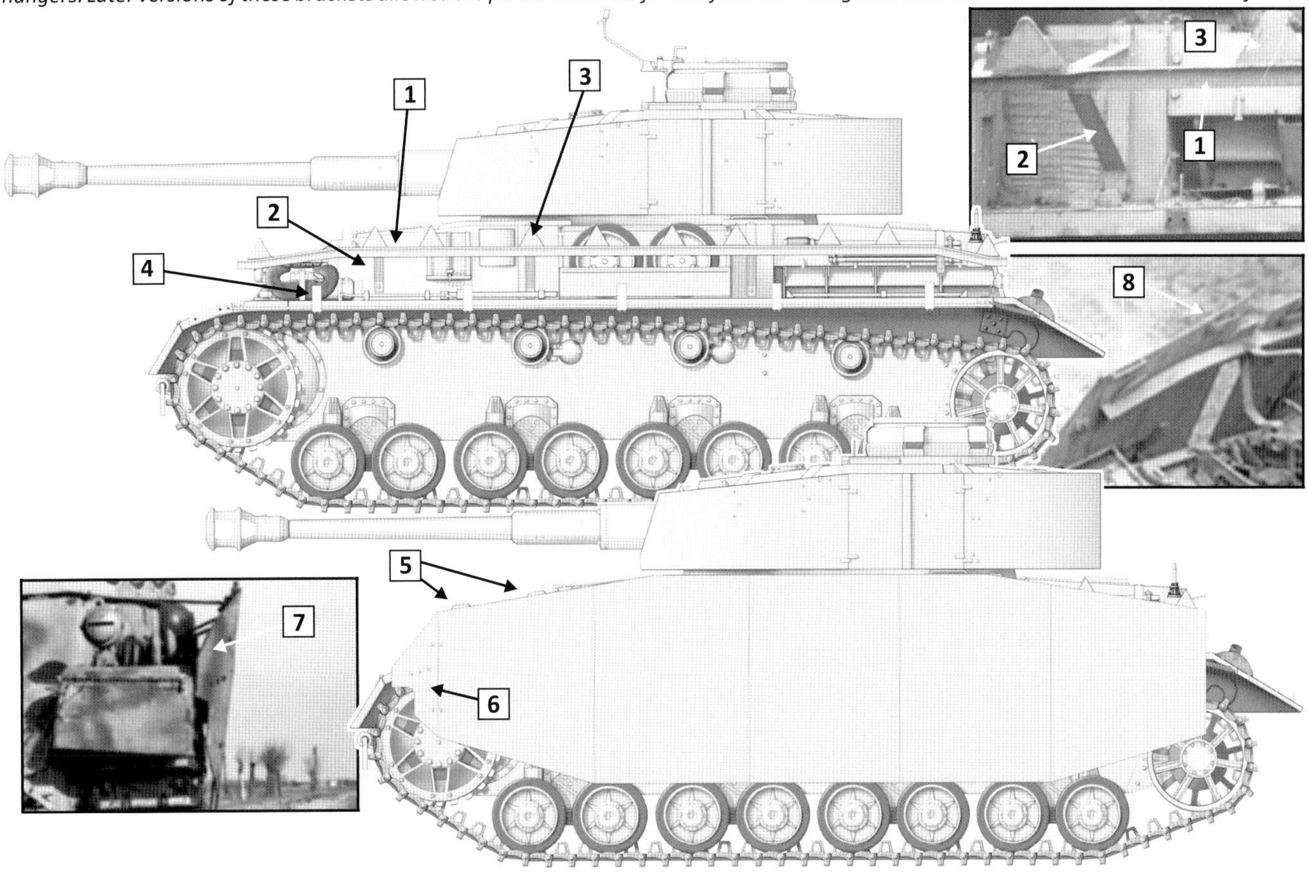

Below: A Pzkpfw IV ausf H of II.Abteilung, Panzer-Regiment 3 photographed just prior to the Normandy battles fitted with the Series II Schürzen. The cut-out on the lower edge of the forward plate (8) was an identifying feature of Nibelungenwerk-assembled vehicles and that factory also mounted the side plates of the turret Schürzen 50mm further forward which slightly changed the layout of the access doors. The notations on the hull Schürzen (9), indicating on which side and in what order they should be placed, was a common affectation within this battalion. Note that the forward triangular hangers (10) have been cut down to follow the line of the hull Schürzen. Note also how the brackets for the hull Schürzen are fixed in place on the superstructure roof by a single bolt (11).

Although the Series II Schürzen solved a number of the problems experienced with the Series I design, the concerns over the build-up of dust and fumes between the armour plates and the hull side remained. In addition, the hull Schürzen added a considerable amount of weight to the vehicle and as early as May 1943 experiments were being conducted where the metal plates were replaced by 5mm and 6mm-thick wire mesh. Having established that this material, which was already in production, could absorb the impact of a 14.5mm anti-tank round work progressed on a plan to apply this type of armour, referred to as Drahtgeflechtsschürzen, to all Panzers. A new supporting rail (1) was built from tubular steel and was held to the hull by four brackets (2) which were bolted to the hull in a similar fashion to the earlier designs. The two central brackets were also fixed to the trackguard (3) to provide extra stability. The hangers on the trackguard (4) were also redesigned and could be raised or lowered to alter the angle of the Schürzen to accommodate the 568mm-wide Ostketten tracks. The mesh screens were fitted with metal frames (5) to prevent damage to the edges. The arrangement was tested on a Pzkpfw IV ausf H in late 1943 but it was not introduced into production until September 1944 and was first fitted to Pzkpfw IV ausf J Fgst Nr. 92301.

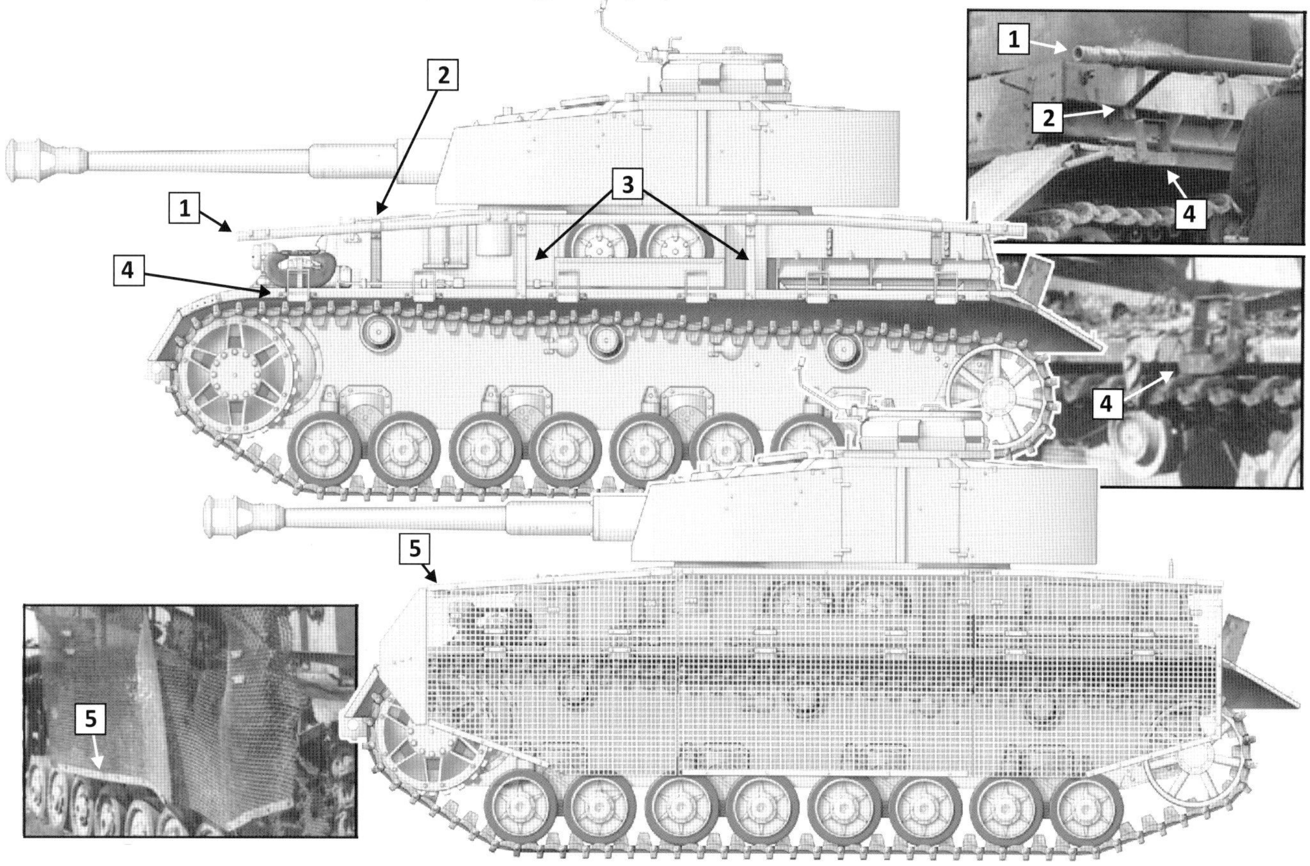

Below: A Pzkpfw IV ausf J photographed in May 1945 near the town of Namest nad Oslavou, west of Brno in the present-day Czech Republic. This vehicle may be Fgst Nr.110089, meaning that it was assembled at the Nibelungenwerk in February 1945. Note the large towing bracket on the hull rear (6) and the Flammentöter exhaust mufflers (7) which are the shorter, larger-diameter versions. The Drahtgeflechtsschürzen is largely undamaged and is only missing the small front plate which was still made from 5mm-thick steel. Note the frame of the mesh screen (8) which provided a degree of support and also prevented the edges of the mesh from snagging on obstacles. The rectangular metal sections visible here (9) were bolted to simple locks which were fitted into the trackguard hangers. Interestingly, it was also felt that the wire mesh provided a degree of disruptive camouflage by breaking up the general outline of the tank.

PZKPFW IV PRODUCTION, FEBRUARY 1944-APRIL 1945

The chart below shows the production figures for the Pzkpfw IV tank from February 1944, when the first Pzkpfw IV ausf J left the assembly lines, until the Nibelungenwerk (NiWerk) was overrun by units of the Red Army in April 1945

Delays in manufacture were caused by the Allied bombing campaign but a more damaging aspect may have been the almost constant amendments to production orders often, but not always, at the behest of Hitler, whose meddling in technical matters is legendary. As early as April 1943 the engineers at Krupp were informed that production of the Pzkpfw IV was to be adandoned in favour of the Panther I and Panther II but in August of the same year this order was reversed and Krupp was instructed to resume assembly of the Pzkpfw IV with a production target of 150 stipulated by the HWA. Before the month was out the quantity had been altered to 100 and by December 1943 production of the Pzkpfw IV was gradually replaced by the Sturmgeschütz IV assault gun. This was initially a short-term solution intended to alleviate the delays caused by the damage caused by Allied air raids on the Alkett factory which built the Sturmgeschütz III. But by January 1944 Krupp had ceased production of the Pzkpfw IV completely, assembling just thirty tanks in that month. Similarly, Vomag gradually shifted production to the Jagdanzer IV tank destroyer and by June, after assembling about 182 Pzkpfw IV ausf J models, ceased production of the tank altogether.

Although assembly continued at NiWerk, that company was also affected by requests for production changes and the proposed introduction of new designs. In July 1944 Hitler ordered that Pzkpfw IV production should cease immediately to be replaced by the Panzer IV/70(A) tank destroyer and ultimately the Panzer III/IV, a design which never progressed beyond the drawing stage. Both would eventually be replaced by the planned Panther II. The development of the Jagdpanzer IV, Panzer IV/70(A) and Panzer IV/70(V) tank destroyers is covered in TankCraft 26: Jagdpanzer IV, German Army and Waffen-SS Tank Destroyers, Western Front, 1944-1945.

Between February 1944, when the first Pzkpfw IV ausf J, left the assembly lines, and April 1945 a total of 3,160 gun tanks, purpose-built command vehicles and specialist artillery observation tanks were manufactured and accepted by the HWA. That number, and the figures given below, can be regarded as accurate but it should be remembered that many records for the last weeks of the war have not survived and in fact NiWerk claimed that an additional fifteen new Pzkpfw IV tanks were built in April 1945. It is certain that another twenty-eight were assembled after May 1945 while the factory was under Soviet control. The Russians appropriated the facilities of Truppenübungsplatz Milowitz, modern-day Milovice north of Prague, and approximately 245 Pzkpfw IV tanks passed through here with Czechoslovak crews before the type was replaced. In April 1955 the Syrian government purchased forty-five of these tanks, for the princely sum of £4,500 each, and they last saw service on the Golan Heights in 1967.

Month	Manufacturer		Remarks
	Vomag	Nibelungenwerk	
February 1944	54 (Pzkpfw IV ausf H)	100 (Pzkpfw IV ausf H)	19 in repair (1)
		100 (Pzkpfw IV ausf J)	
March 1944	60 (Pzkpfw IV ausf H)	250 (Pzkpfw IV ausf H)	18 in repair
			The process of converting rebuilt or refurbished
			tanks as Panzerbefehlswagen IV command vehicles
			began in this month with a total of 88 conversions
			completed by July 1944
April 1944	40 (Pzkpfw IV ausf J)	259 (Pzkpfw IV ausf J)	21 in repair
			10 converted to Artillerie-Beobachtungspanzer IV (2)
May 1944	28 (Pzkpfw IV ausf J)	274 (Pzkpfw IV ausf J)	28 in repair
			22 converted to Artillerie-Beobachtungspanzer IV
June 1944		300 (Pzkpfw IV ausf J)	30 in repair
			13 converted to Artillerie-Beobachtungspanzer IV
July 1944		300 (Pzkpfw IV ausf J)	26 in repair
			13 converted to Artillerie-Beobachtungspanzer IV
August 1944		300 (Pzkpfw IV ausf J)	19 in repair
			25 converted to Artillerie-Beobachtungspanzer IV
			During August and September a total of 17 new
			tanks were converted to Panzerbefehlswagen IV
September 1944		180 (Pzkpfw IV ausf J)	20 in repair
			25 converted to Artillerie-Beobachtungspanzer IV
October 1944		187 (Pzkpfw IV ausf J)	14 in repair
			25 converted to Artillerie-Beobachtungspanzer IV
November 1944		200 (Pzkpfw IV ausf J)	8 in repair
			20 converted to Artillerie-Beobachtungspanzer IV
December 1944		195 (Pzkpfw IV ausf J)	5 in repair
			25 converted to Artillerie-Beobachtungspanzer IV
January 1945		170 (Pzkpfw IV ausf J)	4 in repair
February 1945		160 (Pzkpfw IV ausf J)	1 in repair
			25 converted to Artillerie-Beobachtungspanzer IV
March 1945		55 (Pzkpfw IV ausf J)	3 in repair
April 1945		50 (Pzkpfw IV ausf J)	Provisional figure only (3)

1. The figure given here for vehicles in repair is taken from the Führungsliste for February 1944 and does not actually effect the number manufactured or accepted during that time.

2. Pzkpfw IV tanks were taken directly from the assembly line for conversion to what was referred to as Artillerie-Beobachtungswagen IV from April 1944 when the first ten conversions mentioned here were completed. But it was some time before any vehicles were actually accepted by the HWA inspectors and it was not until July 1944 that the first tanks were ready for assignment to operational units.

3. As mentioned above, a total of twenty-eight vehicles were assembled from parts at NiWerk in May 1945 under Soviet control.

PANZER-REGIMENT, APRIL 1944

All formations of the German Army were organised according to detailed tables of establishment referred to as Kriegsstärkenachweisung (KstN) which showed the official composition of a unit, listing the exact number of personnel and type of equipment from vehicles to small arms. The tables were numbered and dated, the latter being all-important as many numbers were duplicated. The KstN were accompanied by Kriegsausrüstungsnachweisung (KAN) which listed minor items in detail such as tools and fuel containers but these need not concern us here. Efforts to simplify the organisation of Panzer units began as early as December 1943 and by April of the following year instructions had been issued to institute the so-called frei Gliederung, or free organisation, concept where support and transport elements were taken from companies and concentrated at battalion or regimental level. New KstN were introduced which were to take effect from 1 April 1944 for units under the control of the Inspekteur der Panzertruppen and Waffen-SS regiments were ordered to adopt the changes in May. Although the April changes were examined in some detail in TankCraft 35: Panzer IV Medium Tank, German Army and Waffen-SS, Normandy Campaign, Summer 1944, *I have included this chart to allow the reader to make a comparison with the establishment introduced in November 1944 which is shown on the facing page.*

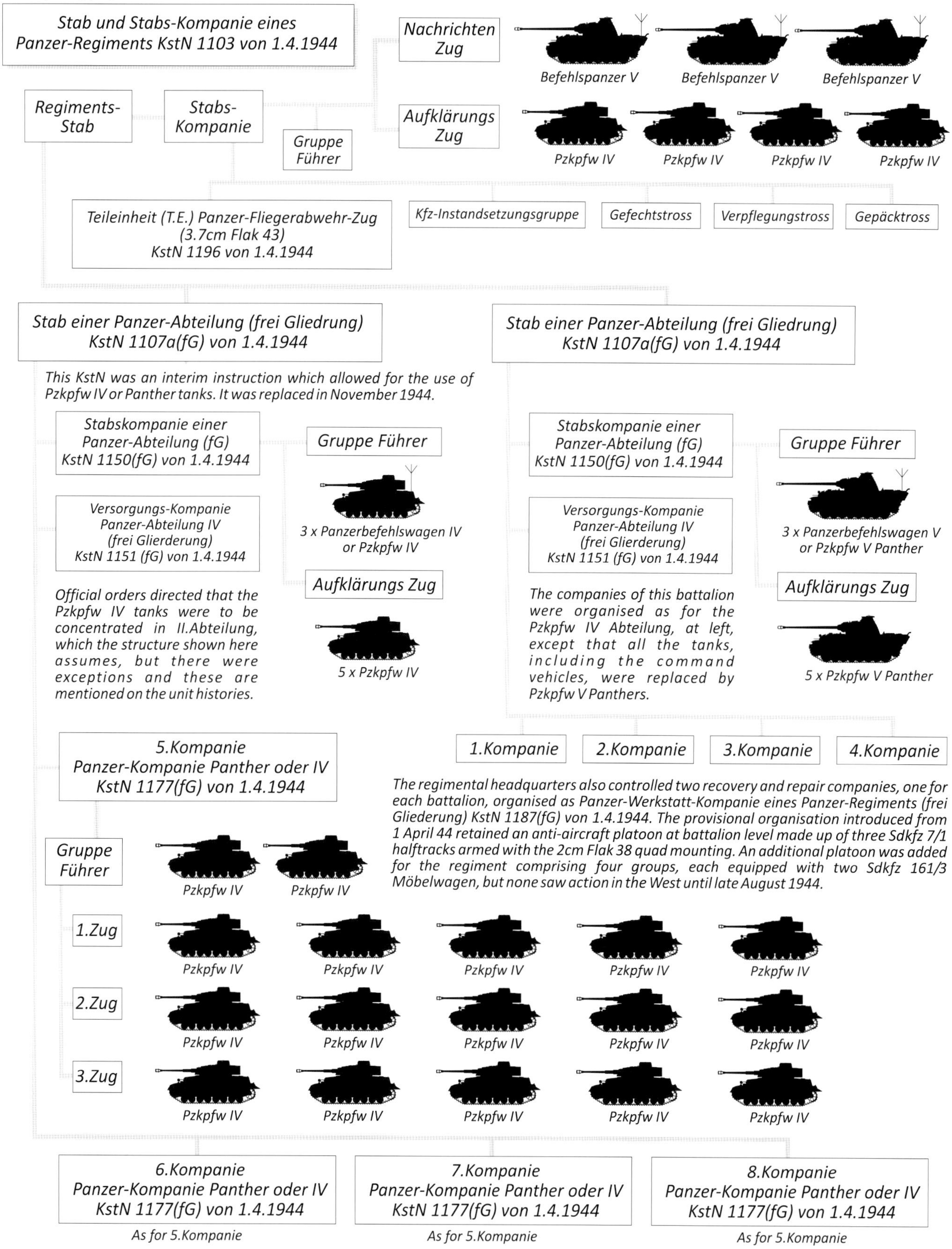

PANZER-REGIMENT, NOVEMBER 1944

On 3 August 1944 Oberkommando des Heeres (OKH), the high command of the army, issued orders that all Army Panzer divisions were to be reorganised according to a standardised configuration referred to as Gliederung Panzer-Division 44. This was to be accomplished using the orders issued for 1 April 1944 and a number of provisional changes were also implemented. In November 1944 further instructions were issued to be used for new formations or units undergoing rebuilding and these allowed for companies to be equipped with either the Panther, Pzkpfw IV tank or Panzer IV/70 tank destroyer. The companies could be made up of either seventeen, fourteen or ten vehicles, with the three platoons containing five, four or three tanks respectively. The regimental headquarters was reorganised and actually gained a number of Kettenkrad halftrack motorcyles and Volkswagen light cars but the number of tanks remained the same. The Kfz-Instandsetzung, or vehicle repair, section was expanded to two groups but the addition of Gepanzert, or armoured, to the title was a name change only. The Gefechtstross platoon, which was responsible for weapon and ammunition resupply, the Verpfelgungstross, which dealt with rations, and the Gepäcktross, or baggage train, changed little. It is not known how many formations actually used the November instructions and it is almost certain that none were put into effect until 1945.

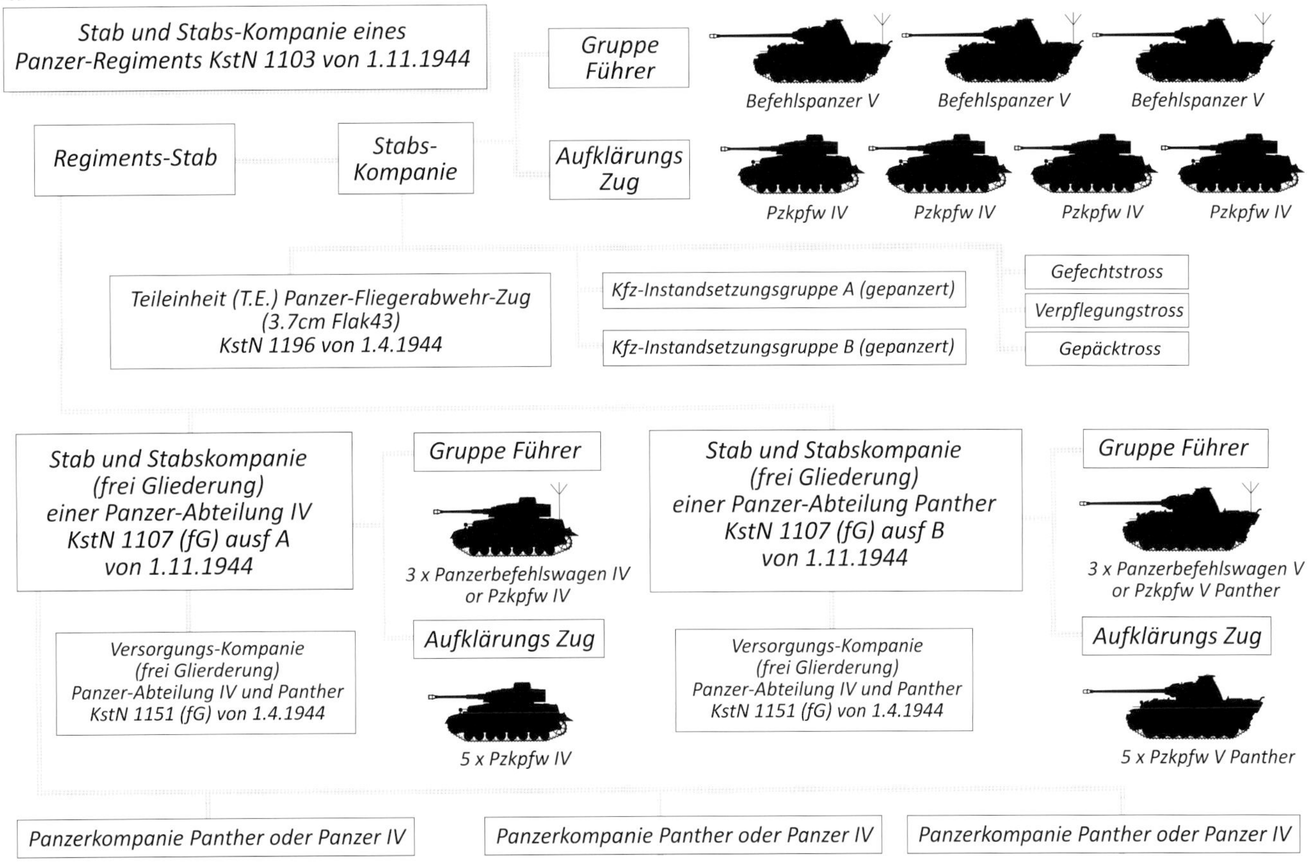

Taking into account the regiment's assets and depending on which vehicles were available, the companies were to be organised using one of the establishments shown below.

PANZER-REGIMENT, APRIL 1945

On 25 March 1945, Oberkommando des Heeres (OKH) ordered that all Panzer and Panzergrenadier divisions, with the single exception of 232.Panzer-Division, were now to be organised in accordance with the instructions issued for Gliederung der Panzer-Division 45. Each division was to contain a single Panzer battalion of four companies which was combined with a Panzergrenadier battalion under a regimental headquarters to form a Gemischte, or mixed, regiment. Detailed KstN and KAN were issued on 1 April 1945 and, as with the November 1944 organisation, the Panzer companies could be filled with either Pzkpfw IV or Pzkpfw V Panther tanks or Panzer IV L/70(A) tank destroyers, referred to in the March order as Panzerwagen 604/9 (7.5cm 42 L/70). Very few of the latter were ever allocated to Panzer companies and, in addition, it should be remembered that at the time the March 1945 order was issued there were less than seventy operational Pzkpfw IV and Panther tanks on the entire Western Front. The number of authorised armoured vehicles, twenty-two Panthers and between twenty and twenty-two Pzkpfw IV tanks or tank destroyers, and the detailed Gliederung which accompanied both the November 1944 and March 1945 orders, suggests the organisation given below. All formations were to report their progress by 1 May 1945 although it appears that none did.

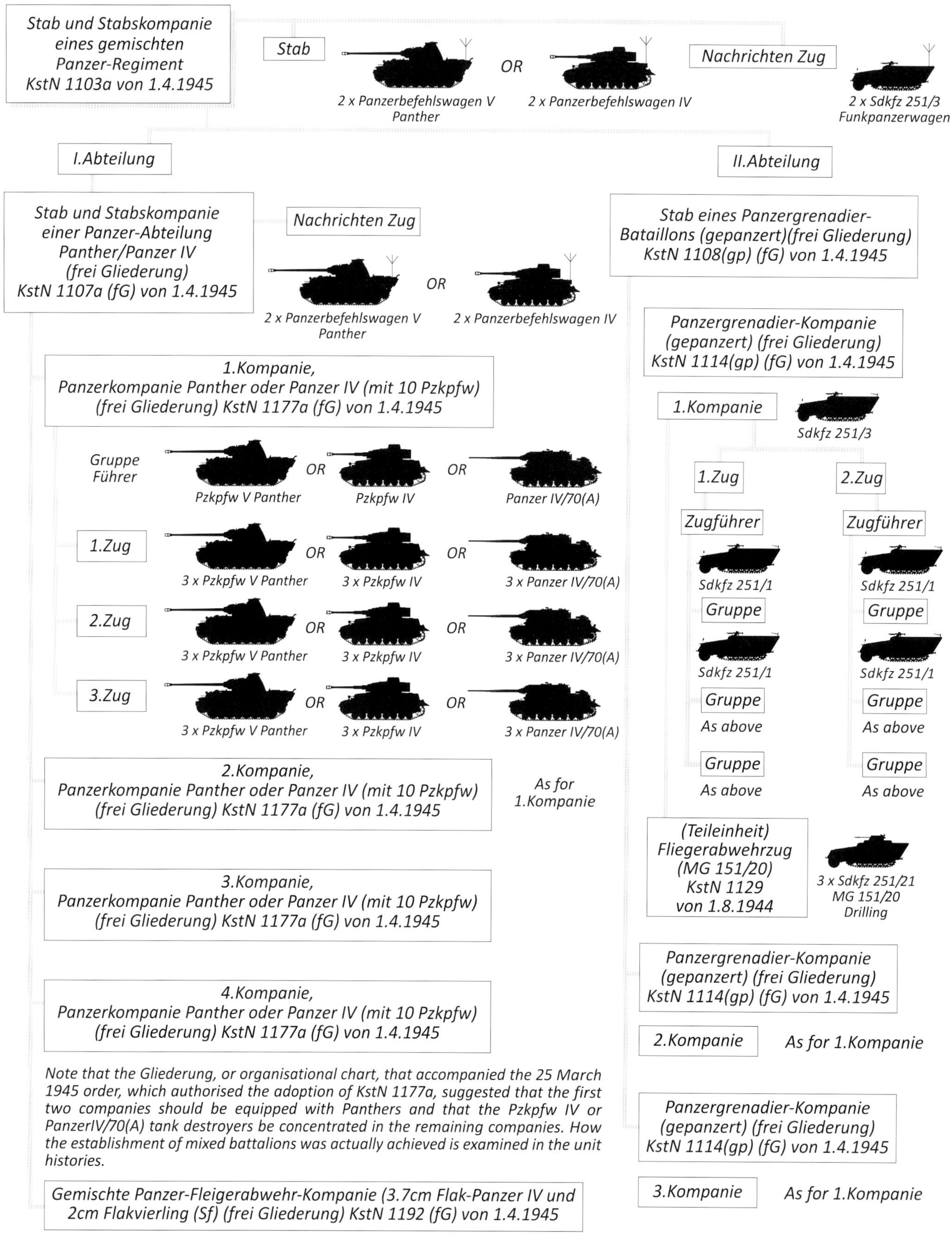

Note that the Gliederung, or organisational chart, that accompanied the 25 March 1945 order, which authorised the adoption of KstN 1177a, suggested that the first two companies should be equipped with Panthers and that the Pzkpfw IV or PanzerIV/70(A) tank destroyers be concentrated in the remaining companies. How the establishment of mixed battalions was actually achieved is examined in the unit histories.

PANZER-DIVISION CLAUSEWITZ, APRIL 1945

The last armoured division raised by the German Army during the war, this formation was created by an order dated Wednesday, 4 April 1945, with the stipulation that the division was to be ready for operations by the following Sunday. The division was to be formed from replacement and training units with the anti-tank battalion ofPanzergrenadier-Division Grossdeutschland and elements of 325.Infanterie-Division. But the infantry units could not be spared from the front and Panzerjäger-Abteilung Grossdeutschland had lost all its tank destroyers in the fighting in the East and had to be refitted with Sturmgeschütz III assault guns before it could join the division. Panzer-Division Clausewitz was eventually raised from the remnants of various units including Panzer-Division Holstein, another formation which was a division in name only, and parts of 233.Reserve-Panzer-Division. But the division's real striking power was concentrated in its two armoured battle groups, shown below. Although these groups lacked transport and a full complement of armoured vehicles, most of the tank crews were combat veterans or instructors. Both underwent several organisational changes and were these have been confirmed I have mentioned them.

Panzer-Regiment Clausewitz. 6 April 1945

Formed on 6 April 1945 from the remnants of Panzer-Brigade 106 Feldhernhalle, with twenty-five tanks of different types and elements of 1.Kompanie, Panzerjäger-Abteilung 93.

Gemischtes Panzer-Regiment Clausewitz. 9 April 1945

The regiment's title as shown here was first used in the 9 April 1945 report.

- Stab, Panzer-Brigade 106 — Begleit-Kompanie
- Panzer-Abteilung 2106 — Stab und Begleit-Kompanie
  - Panzer-Kompanie
  - Panzer-Kompanie
  - Versorgungs-Kompanie

Pzkpfw V Panther — Jagdpanzer IV/70

Exact figures for the Panzer companies are not available

- II.Abteilung (gemischtes), Panzer-Regiment Feldherrnhalle — Stabskompanie
  - Panzergrenadier-Kompanie (gep)
  - Panzergrenadier-Kompanie (gep)
  - Panzergrenadier-Kompanie (gep)
  - Versorgungs-Kompanie

II.Abteilung was reformed in March 1945 as a Panzergrenadier unit after the original Panzer battalion had been wiped out on the Eastern Front.

Panzer-Abteilung Putlos. 12 April, 1945

- Stabskompanie
  - Panzer-Kompanie (gemischtes)
  - Panzer-Kompanie (gemischtes)
  - Panzergrenadier Kompanie (gep)
  - Panzergrenadier Kompanie (gep)
  - Versorgungs Kompanie (mot)

It is likely that the 12 April 1945 establishment was merely the proposed strength of the battalion and was never realised.

Panzer-Abteilung Putlos. 14 April, 1945

- Stabskompanie
- 1. (Schwere) Panzer-Kompanie

2 x Pzkpfw VI Tiger — 10 x Pzkpfw V Panther

- 2. (Gemischtes) Panzer-Kompanie

7 x Pzkpfw IV — 4 x Jagdpanzer IV/70

1 x Jagdpanzer — 1 x Stug III

- Panzergrenadier-Kompanie (gep)

Some sources state the battalion had two Panzergrenadier companys.

On the afternoon of 17 April the battalion was re-organised as Kampfgruppe Benningsen and the figures given for that day are shown below.

Gemischtes Panzer-Regiment Clausewitz. 17 April 1945

- Stab, Panzer-Brigade 106 — Begleit-Kompanie
- Panzer-Abteilung 2106 — Stab und Begleit-Kompanie
  - Panzer-Kompanie
  - Versorgungs-Kompanie
  - Panzer-Kompanie (gemischtes)

StuG IV

Although the report for 17 April 1945 includes the StuG IV company, an additional notation is made that it was in Dresden on 15 April 1945 and it is not mentioned again, suggesting that it never reached the battalion.

Pzkpfw V Panther

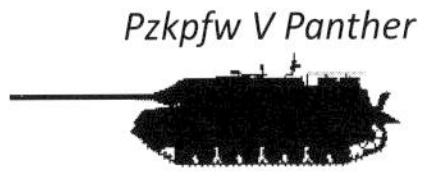

Jagdpanzer IV/70

- II.Abteilung (gemischtes), Panzer-Regiment Feldherrnhalle — Stabskompanie
  - Panzergrenadier-Kompanie (gep)
  - Panzergrenadier-Kompanie (gep)
  - Panzergrenadier-Kompanie (gep)
  - Versorgungs-Kompanie

The report for 28 April 1945 gives a regimental headquarters, one mixed Panzer-Kompanie, a single Panzergrenadier-Kompanie and the Versorgungs-Kompanie.

Kampfgruppe von Benningsen. 17 April 1945

- Stabskompanie
- Panzer-Kompanie

2 x Pzkpfw VI Tiger — 12 x Pzkpfw V Panther

- Panzer-Kompanie

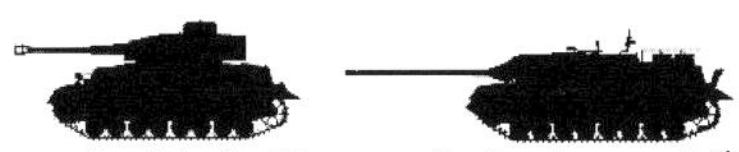

7 x Pzkpfw IV — 4 x Jagdpanzer IV/70

1 x Jagdpanzer — 1 x Stug III

- Panzergrenadier-Kompanie (gep)

- Versorgungs Kompanie (mot)

Major von Benningsen's battle group also reported that nine Sdkfz 250 and ten Sdkfz 251 halftracks were on hand although these may not have all been allocated to the Panzergrenadier company. In addition, this unit contained two 15cm self-propelled infantry guns and three towed anti-tank guns.

Dragon Models Ltd
B1-10/F., 603-609 Castle Peak Rd,
Kong Nam Industrial Building,
Tsuen Wan, N. T., Hong Kong
www.dragon-models.com

Tamiya Inc
Shizuoka City, Japan
www.tamiya.com

Trumpeter/Hobby Boss
NanLong Industrial Park, SanXiang,
Zhong Shan, GuangDong, China
www.trumpeter-china.com
www.hobbyboss.com

Academy Plastic Models
521-1, Yonghyeon-dong, Uijeongbu-si,
Gyeonggi-do, Korea
www.academy.co.kr

Hobby Fan/ AFV Club
6F., No.183, Sec. 1, Datong Rd, Xizhi City,
Taipei County 221, Taiwan
www.hobbyfan.com

Zvezda (Zvezda-America)
www.zvezda-usa.com
Note that the Russian catalogue is not the same as the US version.

Italeri S.p.A.
via Pradazzo 6/b,
40012 Calderara di Reno, Bologna, Italy
www.italeri.com

Takom
www.takom-world.com

Border Model
Xizhang Industrial Park, Huishan District,
Wuxi City, P.R. China
www.bordermodel.com

Sturmpanzer.com
www.sturmpanzer.com

Rye Field Models
www.ryefield-model.com
An almost non-existent website. I would recommend one of the on-line retailers.

Hauler
Jan Sobotka,
Moravská 38, 620 00 Brno,
Czech Republic
www.hauler.cz

Voyager
Room 501, No.411 4th Village,
SPC Jinshan District, Shanghai 200540
P.R.China
www.voyagermodel.com

Griffon Model
Suite 501, Bldg 01, 418 Middle Longpan Rd,
Nanjing, P.R. China
www.griffonmodel.com

Aber
ul. Jalowcowa 15, 40-750 Katowice, Poland
www.aber.net.pl

E.T. Model
www.etmodeller.com

Friulmodel
H 8142. Urhida, Nefelejcs u. 2., Hungary
www.friulmodel.hu

Modelkasten
Chiyoda-ku Kanda, Nishiki-Cho 1-7, Tokyo,
Japan
www.modelkasten.com
Very difficult to navigate but worthwhile.

Theodoros Kalamatas Modelling Worshop
www.facebook.com/Theodoros-Kalamatas-Modelling-Workshop

Eduard Model Accessories
Mirova 170, 435 21 Obrnice,
Czech Republic
www.eduard.com

Master Club
www.masterclub.ru
It appears that this firm is closely associated with Armour35, a Russian mail-order firm.

Model Artisan Mori
Yasutsugu Mori,
Maison Suiryu 302, Kunoshiro-cho 1-10,
Yokkaichi-City, Mie 510-0072, Japan
www.artisanmori.web.fc2.com

Model Factory Hiro
Yubinbango 121-0063 Adachi-ku, Tokyo,
Higashihokima 2-Chome, 3-8 Japan
www.modelfactoryhiro.com

RB Model
Powstancow Wlkp.29B,
64-360 Zbaszyn,
Poland
www.rbmodel.com

M Workshop Singapore
91 Bencoolen St, Sunshine Plaza 01-58,
Singapore
www.themworkshop.com

ROCHM Model
www.rochmmodel.com
rochmmodel@gmail.com
This company has a huge selection of parts and accessories including a 1/35 scale Orterkompass 38.

Royal Model
Via E. Montale, 19-95030 Pedara, Italy
www.royalmodel.com

In writing this book I referred extensively to the *Panzertracts* series of booklets created by the late Thomas Jentz and Hilary Doyle and to a lesser extent, the Pzkpfw IV studies of Waldemar Trojca. Another invaluable resource is Thomas Jentz's *Panzertruppen* books which I would recommend to any reader with an interest in German armour. Although dated, *Panzer IV und seine Abarten* by Walter Spielberger, which has also been published in English, was still helpful. The information on unit allocations was all taken from original Zufürung, or supply lists, compiled by the Heereszeugamt between May 1943 and April 1945. Copies of these lists, and many other original documents, can be obtained quite inexpensively from Sturmpanzer.com and I would urge readers to avail themselves of this service. I also relied heavily on the work of Martin Block and the late Ron Klages whose research on vehicle allocations, including those which actually arrived at their destination, would fill many volumes. I would also like to thank Lisa Hooson and Stephen Chumbley, my editors at Pen & Sword. As always, I am indebted to Karl Berne, Valeri Polokov and J.Howard Parker for their invaluable assistance with the photographs and period insignia.

A early model Pzkpfw III and a Pzkpfw IV ausf G tank photographed in the repair workshop of Panzer-Ersatz und Ausbildungs -Abteilung 35 at Bamberg, about 50 kilometres north of Nuremberg. A number of the older tanks, along with elements of 17SS.Panzergrenadier-Division, took part in the defence of the town in April 1945.